QUICKSILVER MINING

MINING

IN THE
TERLINGUA AREA

KATHRYN B. WALKER

QUICKSILVER
MINING
IN THE
TERLINGUA AREA

KATHRYN B. WALKER

Halo
PUBLISHING
INTERNATIONAL

Halo
PUBLISHING
INTERNATIONAL

Halo Publishing International
8000 W Interstate 10, #600
San Antonio, Texas 78230

First Edition, January 2023
Printed in the United States of America
ISBN: 978-1-63765-310-4
Library of Congress Control Number: 2022916797

Halo Publishing International is a self-publishing company that publishes adult fiction and non-fiction, children's literature, self-help, spiritual, and faith-based books. We continually strive to help authors reach their publishing goals and provide many different services that help them do so. We do not publish books that are deemed to be politically, religiously, or socially disrespectful, or books that are sexually provocative, including erotica. Halo reserves the right to refuse publication of any manuscript if it is deemed not to be in line with our principles. Do you have a book idea you would like us to consider publishing? Please visit www.halopublishing.com for more information.

Disclaimer
The conversations in the book all come from the author's recollections, though they are not written to represent word-for-word transcripts. Rather, the author has retold them in a way that evokes the feeling and meaning of what was said, and in all instances, the essence of the dialogue is accurate.

In loving memory of our mother, Kathryn B. Walker

QUICKSILVER MINING IN THE TERLINGUA AREA

Approved:

Frances M. Phillips

Clifford B. Casey

Bevington Reed

Approved:

Bevington Reed
Chairman of the Graduate Council

QUICKSILVER MINING IN THE TERLINGUA AREA

———————————

A Thesis

Presented to the Graduate Council

Sul Ross State College

Alpine, Texas

———————————

In Partial Fulfillment

of the Requirements for the Degree

Master of Arts

———————————

by

Kathryn B. Walker

August 1960

CONTENTS

Preface   13

Chapter I

Background of area   15

Chapter II

Socio-economic factors in mining   34

Chapter III

Development of the mines   43

Chapter IV

Chisos mines   56

Chapter V

Passing of the mining period   74

Essay on source material   78

Bibliography

A. Primary sources   79

B. Secondary materials   88

Appendix I   95

Appendix II   99

Appendix III   101

PREFACE

M any facets of the history of the Big Bend area of Texas have been written up for the preservation of names, dates, and other pertinent information. A great deal of technical material on geology and mining has been published. There has been, however, no consolidated general history of the mining activities in the Terlingua area. It has been the purpose of this study to assimilate the available material relative to the earliest history and the development of the quicksilver mining industry there, and to present it in narrative form for the general reader.

Many people have made contributions to the completion of this project. I want to thank Mr. and Mrs. W. D. Burcham, Macario Hinojos, and Mr. and Mrs. Harry Fovargue for talking with me about mine processes and for relating interesting facts about the history of the mines. The Sul Ross State College librarians, Warren Baxley, Dudley Dobie, and Mrs. Lanell Skinner were helpful. Mrs. Skinner deserves especial credit for her untiring efforts in ordering material and tracing leads on sources not in the Sul Ross Library. Mrs. Sara Pugh and Mr. Worth Frazier were very helpful and patient while I was working in the Office of the County Clerk in Alpine. The Trans-Pecos Abstract Office assisted in locating mine claims and instrument references. Mr. Charles King,

Sul Ross geology instructor, was very helpful in checking the geology descriptions used to explain the quicksilver ores and their locations.

Some of the geology terms may vary from the ones used at present because the principal materials used as historical sources were written between 1896 and 1910. The writer takes full responsibility for geological terminology and phrasing.

CHAPTER I

BACKGROUND OF AREA

D eep in the heart of the area encompassed by the big bend of the Rio Grande lies one of the most barren, rocky, and desolate areas of Texas. The physical aspect does not reveal the true value of the riches stored there by nature. This seemingly arid wasteland held its secret from man until the latter part of the nineteenth century. The Indians and Spaniards had passed the buried treasure of mercury until the 1880's when it was first noted by a Mexican sheepherder.[1] The Indians had left their mark of discovery by drawings on the rocks with the red ore.[2]

The land bounded by the Chisos Mountains on the southeast, the semi-circular band of the Rio Grande south and west, Fresno Canyon also on the west and on the north, and a rather natural boundary of Terlingua Creek on the east comprises the Terlingua area. This district is in Brewster County and is about ninety to one hundred miles

---

[1] "Texas Mines," *Southern Industrial and Lumber Review*, March, 1901, cited in *Alpine Avalanche*, May 5, 1901.

[2] O.W. Williams, letter to his children, Alpine, Texas, March 18, 1902.

from Marfa, ninety to one hundred miles from Marathon, and eighty to ninety miles from Alpine.[3] The Rio Grande is about ten to twelve miles from Terlingua. The area is about 104° west longitude and about 29°30′ north latitude.[4] Here lies the country where men have toiled to the extent of exhaustion for the earth's treasure of cinnabar.

The Terlingua area has a very distinct topography caused by beds that are resistant to erosion and others that give way to erosion very readily. Extensive faulting is responsible in part for many of the topographic features. The northerly portion of the uplift has the highest point along the east part of Fresno Canyon. Lower Cretaceous beds, the highest at an elevation slightly less than four thousand feet, make a jagged, saw-like ridge. In the uplifts the area gives no suggestion of the eroded tertiary, while in the valley, because of lower block faults, the strata of Cretaceous age lie directly beneath the Tertiary.[5]

---

[3] L.N. Halbert rode a bicycle to Terlingua with a cyclometer to measure the distance. The total distance he measured was 91 miles. The mileage he gave was 42-1/2 miles to Butcher Knife Ranch; 27-1/2 miles on to Adobe Walls, and 21 miles on to Terlingua. *Alpine Avalanche*, July 27, 1900.

[4] William P. Blake, "Cinnabar in Texas," Transactions of the American Institute of Mining Engineers, XV (1896), 69. The distances given are those of 1896 and are used because the roads have been changed in recent years. The geology terms used in this study are contemporary of the early 1900's. Some terms have been substituted or changed in more recent years because the newer terms give better descriptions of the area. Robert G. Yates and George A. Thompson, Geology and Quicksilver Deposits of the Terlingua District Texas, Geological Survey Professional Paper, 312 (Washington: United States Government Printing Office, 1959).

[5] Benjamin F. Hill, *The Terlingua Quicksilver Deposit, Brewster County, Texas* (Bulletin of University of Texas, No. 15, The University of Teas Mineral Survey, Bulletin No. 4, October, 1902), p. 15.

A very outstanding topographical feature of the Terlingua area is Black Mesa,[6] a northeasterly to southwesterly trending elliptical dome of lower Lower Cretaceous Limestone. Because of erosion, the core of igneous intrusive rock is exposed near the center of the uplift. Other notable features, California Hill and Clay Mountain, which are igneous intrusions in opposition to the white limestone and dark shades of clays and lavas of the district, are very pronounced.[7]

The quicksilver ore has been found in scattered areas of the Chisos, Terlingua, and related region. Included within the wide boundaries ". . . from the Mariscal Mountain on the east to the Lajitas Mesa on the west and from beyond the Mexican boundary on the south to the Christmas Mountains on the north" not all the ore-bearing deposits graded high enough to justify production.[8] The mapping of the structural geology of the Terlingua and Chisos Mountain quadrangle has been completed. The larger deposits of cinnabar are located in vertical fissures. The ore to be found at lower levels is possibly below the water level of the district.[9] The mercury-bearing ore may be found in structures where the impervious cap rocks have already been removed by erosion or in "lower anticlines or domes or sealed faults, blocks, still capped by impervious

---

[6] Hill wrote the name "Block Mesa."

[7] Hill, *op. cit.*, p. 15.

[8] E. H. Sellards and C. L. Baker, *The Geology of Texas, Vol. II: Structural and Economic Geology*, Bulletin 3401 (January 1, 1934), p. 532.

[9] *Ibid.*, p. 553.

rocks."[10] In a discussion on the anticlinical theory in relation to quicksilver, J. A. Udden described the deposits as follows:

The ore in the Marfa and Mariposa mine, and also in the old Terlingua mine, has been deposited mostly in joints, fissures, and cavernous openings that extend down in the upper surface of the Georgetown limestone. The material filling these fissures is locally known as *jaboncillo*. It is a material of mixed nature, consisting in places largely of clay and in others of materials quite like caliche. It is evident that it has been formed in these fissures partly by precipitation from solutions which have followed the lower surface on the Del Rio clay, and partly also by the Del Rio clay itself which has settled perhaps gradually into solution caverns, *pari passu* with enlargement by solution. The *jaboncillo* frequently contains fragments of limestone itself and is in places not unlike a fault breccia, cemented with calcareous material. At the surface, this *jaboncillo* in places changes to caliche, clearly formed at a recent date. Even this caliche contains fragments of cinnabar, which apparently have been entombed quite recently in the formation of superficial caliche. The fissures extend to varying depths and no doubt in some place, or places, join "pipes," most probably along fault lines, through which the quicksilver has ascended. The great number of these ore-bodies gives out in less than twenty feet below the upper surface of the Georgetown limestone and the mining on most of the hills in the sections mentioned has been in shallow pits which stud the land at the present time. In places, ore occurs at

---

[10] *Ibid.*, p. 554.

great depths and some fissures have been found with open vugs[11] set with calcite and gypsum crystals. It would seem altogether likely that further prospecting may result in the finding of deeper pipes and it would seem quite probable that these should, as already stated, be found in some of the faulted fissures known to exist.[12]

Mercury, like gold, is where one finds it, and not all mercury is found in like formations.[13]

The rocks of the Terlingua area may be divided into two classes, sedimentary and igneous. With the exception of a few scattered deposits, the sedimentary rocks originated in a shallow to deep water marine environment. The sedimentary rocks range in age from Cretaceous and Tertiary. The Cretaceous strata are approximately two thousand feet thick in this area, constituting most of the rock outcrops. Rocks of Tertiary age totaling several hundred feet in thickness occur in widely scattered exposures.[14]

---

[11] *Vug.* A rock cavity lined, but incompletely filled, with mineral matter so that a part of the available space remains empty. *Van Nostrands's Science Encyclopedia* (Princeton: D. Van Nostrand Company, Inc., 1958), p. 1784.

[12] Sellards and Baker, *op.cit.*, pp. 540-541, citing J. A. Udden, *The Anticlinical Theory as Applied to Some Quicksilver Deposits:* University of Texas Bulletin 1822, 1918, pp. 14-15. Map, Structural Map of Trans-Pecos, Plate IV, Appendix, p. 6.

[13] Hill, *op. cit.*, p. 54. Table No. I on page 30-31 shows various types of formations in different countries where mercury is mined. The formations identified in the Chisos area are also included in Table II, page 32. Yates and Thompson, *op.cit.*, p. 7.

[14] Yates and Thompson, *op. cit.*, p. 54. For detail of the structural geology of the area see Plate 1, page 33.

Some of the mercury-minerals found in this area are of the most usual kinds – cinnabar; meta-cinnabarite; black sulphide of mercury, in an uncrystallized state; terlinguaite;[15] calomel, which is chloride of mercury; a mixture of mercury and antimony that at the time was not definitely determined.[16] Cinnabar is found in a beautiful ruby-red crystals, occasionally some as long as three-fourths of an inch, closely associated with calcite and native mercury. The crystals are generally needlelike prisms, but sometimes have a thick flat shape. The needlelike prisms are seen only in calcite veins. A great deal of cinnabar is a mass of crystals, sometimes large in size. The crystals vary in color from bright vermillion to dark reddish brown.[17]

---

[15] A new variety, according to Professor S. L. Penfield, who says that terlinguaite is oxychloride of mercury. Hill, op. cit., p. 28. Some characteristics of terlinguaite are "Luster, brilliant adamantine. Color, sulphur-yellow with a slightly greenish tinge, very slowly darkening on exposure to an olive green. Color of powder lemon-yellow, also slowly becoming olive-green. Transparent or nearly so. Hardness between 2 and 3. Brittle or subsectile." H. W. Turner was the person who made the first reference to terlinguaite. "'In addition to cinnabar, mercury occurs in a native form and as a white coating and as *yellow-green crystals*. Professor S. L. Penfield had identified the . . . greenish crystals as an oxychloride of mercury forming a new mineral species for which I have suggested the terlinguaite.;" A. J. Moses, "Eglestonite, Terlinguaite, and Montraydite," New Mercury Minerals form Terlingua, Texas," *American Journal of Science*, XVI (September), 1903, 258-259, citing H. W. Turner, "The Terlingua Quicksilver Mining District, Brewster Co., Texas," *Mining and Scientific Press* (San Francisco), July 21, 1900, p. 259. For further confirmation as to identification of color and characteristics consult W. F. Hillebrand and W. F. Schaller, *The Mercury Minerals from Terlingua Texas* (United States Geological Survey, Bulletin 405, 1909), pp. 7-8; 83-142.

[16] Hill, *op. cit.*, p. 28.

[17] *Ibid.*, p. 29.

Native mercury or "free mercury" may also be found in this field in sizeable amounts. Some calcite veins have been found to yield as many as twenty pounds of "free" mercury. Cream colored clay may also contain some native mercury.[18] Once in a while a Mexican man would find a large rock that would emit free mercury during the hotter part of the day. The man would put some sort of a container under the rock to catch the precious silver thread of mercury as it was yielded from the rock.[19]

Occasionally a man would be using a pick in a shaft and find as much as a cup of free mercury, which will weigh about nine and one-half pounds. Mercury is the only precious metal that is in liquid form after it is refined.[20] This "elusive and beautiful liquid"[21] in pure form was not often found, and when it was, the find was usually a much smaller amount. In the California Hill Mine black calcite[22] may be spotted with "free" mercury. Some of these rocks could be broken and the mercury could be shaken from the rock.[23]

---

[18] *Ibid.*

[19] Macario Hinojos, interview, Terlingua, Texas, March 7, 1959.

[20] C. A. Hawley, *Life Along the Border* (Spokane: Shaw and Border Company, 1955), p. 119; Sellards and Baker, *op. cit.*, p. 511. Hawley wrote of his experiences while working for the Marfa and Mariposa Mining Company and the Chisos Mining Company. Hawley had been a teacher in the North, but had come South for his wife's health.

[21] "Quicksilver," *Alpine Avalanche*, December 21, 1900.

[22] Yates and Thompson, *op. cit.*, p. 66.

[23] Hinojos, interview, Terlingua, Texas, June 22, 1958.

The occurrence of quicksilver ore has been described thus:

. . . in the Marfa and Mariposa mine and in the old Terlingua mine much of the ore has come from brecciated fissures in the upper surface of the Edwards limestone. This was originally directly covered by the next to impervious Del Rio clay, which at the present time is mostly removed. Evidently in this case, also, the ore was precipitated from rising solutions at a level where these were hemmed in by the impervious cover of clay. As cinnabar has been found in this mine also in a porphyry, the ore is in that respect in a situation quite similar to that at Big Bend. If the mineralized solution followed the course of a body of an intrusive in the limestone it would be likely to have a comparatively open and unimpeded passage along this intrusive, as long as this transversed the limestone. For in limestone dikes are quite uniform in their development. By reading the overlying clay the solution would find the dikes less regular and the clays also would hinder its ascension. The solution would naturally follow the upper surface of the limestone and precipitates would find lodgment in its fissures under the clay.[24]

The small amount of plant and animal life found in the southern part of Brewster County was not in balance with the hidden treasure of the soil. Plants and animals were generally sparse and small in stature.

A desert-type vegetation and small wild animals are characteristic of this area. Some of the plants are small cacti of all kinds, ocotillow, greasewood, mesquite, candelilla,

---

[24] Udden, *Sketch of Geology* . . . , pp. 89-90

chino grass, very limited grama grasses, Bermuda grass in the vegas,[25] creosote, mesquite grass, burro grass, sotol and lechuguilla. Most of these plants are found only in arid lands such as the Terlingua district. In the higher parts of the Chisos Mountains, however, are found pine, spruce, fir, cedar, and century plants as well as other plants native to high altitudes.[26]

The small animals of the locale include burros, rabbits, ringtails, lizards, coyotes, javelinas, snakes of many kinds, deer, a few antelope, bobcats, foxes, and polecats, to name some of the more common animals. In the Chisos Mountains larger animals such as the panther and bear are found. The Mexican people depended a great deal on native plants and animals to aid in their existence in this desolate region.[27]

The best time to try to work in Terlingua is early morning or late afternoon, as it is very hot at midday. The Mexicans always take a siesta[28] after lunch and wait until later in the afternoon to finish their work. The winters are mild and short. Seldom does it stay cold for more than two days at a time. Close to Terlingua a great deal of irrigation has produced some very good cotton, cantaloupe and other crops in more recent years. The rainfall is very light;

---

[25] *Vega* means "meadow," or " valley with grass."

[26] Adolph H. Witte, "Notes on the Big Bend Region of Texas," *Bulletin of the Texas Archeological and Paleontological Society*, XV (September, 1943), 108.

[27] See Appendix I, Use of Native Plants.

[28] *Siesta* means "nap" or "rest."

however, a good rain up in the mountains will soon fill all the creeks to overflowing and stop traffic for many hours.[29] The creeks then emptied into the Rio Grande, a river that the Mexicans have given various names.[30]

Terlingua has also had a series of names. The name *Terlingua* has several possible origins. One term used was *Las Linguas*.[31] Another early name was *Tres Lenguas* ("Three Tongues"). Later these words were slurred together as

---

[29] To illustrate the treacherous mountain streams after a rain, Hawley relates the sudden death of one of his new found friends in a rain swollen stream. "Mr. Donaldson met a tragic death shortly after our meeting. He and a subordinate officer hired a Mexican boy to take them over to the nearby silver mining town of Shafter, in Presidio County. It was a very hot day, and they made the trip in a hack, equipped with the usual canvas top and side curtains. A mile or so from Shafter they crossed an arroya [sic], or dry water course. While they were in town a heavy shower took place some miles up the arroya [sic], although no rain at all fell in Shafter. When they reached the arroya [sic] on their way home they were surprised to find it a raging torrent. The Mexican boy did not want to attempt the crossing, but Mr. Donaldson and his companion thought they could make it all right and urged the boy to drive in, which he did. The water was deeper and was coming with much greater force than they realized, and they had gone but a short distance when the hack was overturned and swept down the stream. The Mexican boy jumped out onto the back of one of the mules and by doing so saved his own life, but the two men were drowned. A day or two later their bodies were found in the hack a mile or two below the crossing.

Within a few hours the water was all gone and the crossing as dry as usual. Such cloudbursts, converting dry water courses into dangerous torrents, are common throughout the Big Bend Country of western Texas." Hawley, *op. cit.*, p. 40. Mr. Donaldson was an Internal Revenue man stationed at Lajitas about 1907.

[30] "Rio Bravo" means "Brave River;" "Rio Puetago" local term meaning "Muddy River."

[31] United States Court Western Division of Texas, Decree, May 13, 1901, Deed Records, Office of County Clerk, Brewster County, Alpine, Texas, VIII, 164-165.

Tesslinguas; finally, the name was changed to *Terlingua*.[32] The three tongues referred to the three spoken languages – English, Spanish, and Indian.[33] A different story is that at one time the Apache, Comanches, Shawnees supposedly lived there in peace and constituted the three languages. Another variation for the meaning of Terlingua was in reference to Goat, Calamity, and Crystal Creeks all flowing into one main stream. These creeks were then referred to as the three tongues of Terlingua Creek, which is a tributary of the Rio Grande.[34] Still another possible account for the name Terlingua has been told. In 1912, an eighty-year-old Mexican man who had studied to be a priest, but has never completed his studies because he did not like the priesthood, told another version of the Terlingua name. In going over some very old documents in the church where he was studying, he found mentioned Terlingua Creek and a small Indian village having the same name of Terlingua. The record was hard to read as it was old and faded and the Spanish it was written in was not familiar to him. A different variation is based on a drink made from a particular plant in the area. This intoxicating drink was called *tezlingo*, and supposedly the creek was called the

---

[32] "What Does Terlingua Mean?" reprint from the *San Antonio Express, Alpine Avalanche*, March 23, 1903.

[33] Hinojos, interview, Terlingua, Texas, June 22, 1958.

[34] "Place Names are Significant for the traveler as He Passes Through Texas," *Alpine Avalanche*, Sixtieth Anniversary Edition, September 14, 1951, p. 40.

same thing. Later the word was changed in some manner to "Terlingua."[35]

Closely related to the Terlingua is a group of mountains with Indian and Mexican folklore attached to the name. This range of Mountains is the *Chisos* that is commonly accepted to mean "Ghost" in Spanish. An unusual phenomenon in the mountains occurs after a heavy rain. A queer chemical reaction after a heavy rain will cause small balls of fire to bounce down the mountains, terrifying the superstitious natives, making them believe the spirits had caused it. Thus, the name *Chisos* was given to this group of mountains. Another effective tale is based on the peculiar outline of the mountain peaks against the skyline. The suggestion of a mummy's form is possibly responsible for the title Chisos Mountain.[36]

Not only is the origin of their names uncertain but there is little record of their early people as the district is downstream from the junta,[37] which was the principal route of Spanish exploration. Of course, various tribes raided back and forth across the Rio Grande near Terlingua, but no

---

[35] Paulina Cepeda, "History of Terlingua," brief term paper, Sul Ross State College [1956], citing an unpublished manuscript by A. W. Fulcher, 1949.

[36] Freda Gibson, "Local Place Names," *West Texas Historical and Scientific Society,* Bulletin 21, I, December 1, 1926, pp. 40-41.

[37] *junta*- junction of the Rio Grande and Concho River, a river draining a portion of northern Mexico, at Presidio, Texas.

evidence is shown of any permanent settling in the area until the late 1800's.[38]

Indian traces abound in these vicinities, and the recesses of the adjoining mountains afford secure retreats, where the animals plundered from the Mexican settlements are driven to recruit, in preparation for the passage across the Rio Grande into Texas.[39]

The Indians that traversed the Terlingua area were the first to use the cinnabar ore. The body was frescoed and pictographs drawn on the rocks in the area. Near Comanche Springs in the present Big Bend Park area, the front of a limestone ledge has a pictograph of a buffalo with a curl or twist in his tail that symbolized an angry animal. There are some dots and other vague figures. The

---

[38] For further information of Indian activities, the following references discuss the Indians in various periods along the Rio Grande area. "Texas Prehistory – Indians," *Texas Almanac*, 1857-1957 (*Dallas Morning News*, 1956-57), pp. 44-48. A brief general discussion of Indian tribes in Texas. Rupert Norval Richardson and Carl Coke Rister, *The Greater Southwest* (Glendale; The Arthur H. Clark Company, 1934), pp. 306-458. Mention is made in some instances to the Indians of this area. Carl Coke Rister, *The Southwestern Frontier, 1865-1881* (Cleveland: The Arthur H. Clark Company, 1928). A more local account of Indians. Reports have been made that different exploration groups traversed the Terlingua area. One group approached the area, but crossed into Mexico as the terrain was better and not so hard to travel. Another group coming in from San Antonio began to head north from the Devil's River, across to the Pecos River, west to Wild Rose Pass, then to Presidio del Norte; again missing the Terlingua area. Philip St. George, William Henry Chase Whiting, and Francois Xavier Aubry, *Exploring Southwestern Trails, 1846-1854*, Ralph P. Bieber, editor (Glendale; The Arthur H. Clark, 1938), map, Appendix. W. Turrentine Jackson, *Wagon Roads West* (Los Angeles; University of California Press, 1952), map, p. 38.

[39] C. C. Parry and Arthur Schott, *United States and Mexican Boundary Survey*, Geological Reports (Washington,

D. C.: Cornelius Wendell, 1857), p. 61.

Terlingua area was not buffalo country. The animal possibly represented "Buffalo Tail," a Comanche Chief.[40]

Possibly the first permanent settlement was below the present-day site of Terlingua and nearer the river. A Mexican, Cipriano Hernandez, bought and managed a small trading post near the Rio Grande below the present location of Terlingua. At that time, it did not have a special name; now, the area is referred to as Lower Terlingua. This store made trading possible between Mexico and Texas at that point.[41]

The Rio Grande, which serves as a boundary between Mexico and the United States, was not guarded by either nation until a need arose to justify the guards.[42] The Mexican people of the area crossed the Rio Grande as they traversed the land at will. The finding of the ore at Terlingua is credited to one of these wanderers. This seems to be the most reliable and acceptable explanation for the finding of the ore.

According to this story, Juan Acosta, a Mexican deer hunter and sheepherder, found a piece of cinnabar ore, and was curious about the unusual heaviness of the rock.

---

[40] O. W. Williams, letter to his children, Alpine, Texas, March 18, 1902.

[41] Hinojos, interview, Terlingua, Texas, March 7, 1959.

[42] This study will not discuss the Indian, Spanish, Mexican, and American relations further. A more detailed account of their activities can be found in the following publications; H. S. Thrall, *History of Texas* (St. Louis; N. D. Thompson, 1879), pp. 448-450; Richardson and Rister, op. cit., Hawley, op. cit., pp. 39-40, 81-82. In the Commissioners Court Minutes, Brewster County, I, 409, June 14, 1899, a discussion of the need of more law enforcement along the border is recorded.

Juan felt sure this rock must have some valuable mineral that made it so heavy. The story seems to vary here. One account reports that Juan and some other men planned to investigate further his findings.[43] Another version is that Juan took the ore to Shafter for an assay. The ore was tested and showed to be high grade cinnabar. Some Californians heard of the find and made an extensive search of the area. The men found nothing of any real value in the described vicinity. As they left the Chisos range, a man carved "California Hill" on the face of a rock. When operation of the Marfa and Mariposa Mine started in 1896, the inscription was found on the hill and as a result the hill still carries that name.[44] The Californians were on the largest bed of ore in the Terlingua area and did not find it.

Another "authentic" story of the discovery of the quicksilver is credited to two other men. These two men, George W. Wanless of Jimenez, Chihuahua, Mexico, and Charles Allen of Socorro, New Mexico, heard that Mexicans had found some rich cinnabar ore. The men went to the region and located the deposit.[45] How it was found really does not matter. The ore was there and many people had passed it unknowingly. An industry that affected the lives of many individuals was born in the most southerly portion of Brewster County.

---

[43] "Texas Mines," *Alpine Avalanche*, May 5, 1901.

[44] "Quicksilver First Found in Brewster County by a Mexican Hunter," *ibid.*, September 14, 1951.

[45] "Brewster County Minerals," *ibid.*, March 14, 1902.

TABLE I

COMPARISON OF FORMATIONS, ASSOCIATED ROCKS AND ASSOCIATED MINERALS[46]

Country	Geological Formations	Associated Rocks	Associated Minerals
Austria	Upper Triassic	Dolomite, Sandstone Schists, Slates	Quartz; feldspar; mica; hornblende; pyrite; epsomite; copperas; gypsum; idriolite; graphite; anthracite; bitumen; calcium phosphate; fluorspar; marcasite; barite, calomel
Italy	EoceneVarious (Tertiary) Cretaceous	Various lavas, such as trachyte, andesite, rhyolite Marly limestones, and clays	Bitumen; free sulphur; realger, pyrite; marcasite; quarts; calcispar; gypsum
Mexico	Cretaceous Jurassic	Porphyry, Limestones, Slate	Silver and antimony ores; gypsum; calcspar; fluorspar; free sulphur; quartz, calomel

[46] Hill, *op. cit.*, p. 54

Russia	Carboniferous	Sandstone, Quartzite	Stibnite; pyrite; calcspar
Spain	Upper Silurian and Devonian	Slates, Limestone, Shales, Serpentine, Sandstones, Slates, Rhyolite, andesite, basalt	Calcspar; pyrite; galena; quartz; barite; arsenical pyrite
United States	Early Cretaceous or late Jurassic. In the Tertiary, and in Quarternary Alluvium	Granitic detritus, Limestone, Shales, Serpertine, Sandstone, Slates, Rhyolite, andesite, basalt	Calcspar, pyrite; barite; quartz; gypsum, borax; stibnite; free sulphur mispickel; chalco-pyrite; bitumen; marcasite; millerite; gold and silver ores; fluorspar; copiapite; knoxvillite; redingtonite; calomel; terlinguaite

TABLE II

STRATIGRAPHIC NAMES IN
THE TERLINGUA DISTRICT[47]

			Udden (1907a, p. 21-60)	Sellards, Adkins, and Plummer (1933)	Names used in this report	
Upper Cretaceous		Rocks of Tertiary age	Surface flows	Volcanic rocks	Chisos volcanics	
			Burro gravel and tuff			
	Gulf series	Navarro Group	Crown conglomerate	Crown conglomerate		
			Chisos beds	Chisos beds		
			Tornillo clays	Tornillo formation	Tornillo clay	
		Taylor group	Rattlesnake beds	Aguja formation	Aguja formation	
			Terlingua beds	Taylor formation	Terlingua clay	
		Austin group		Terlingua formation *(restricted)*	Boquillas flags	Upper member
		Eagle Ford group	Boquillas flags	Boquillas flags		Lower member
Lower Cretaceous	Comanche Series	Washita group	Buda limestone	Buda limestone	Buda limestone	
			Del Rio clay	Grayson formation	Grayson formation	
		Fredericksburg group	Not differentiated in the Terlingua district	Georgetown limestone	Devil's River Limestone	
				Edwards limestone		

---

[47] Yates and Thompson, *op.cit.*, p. 7

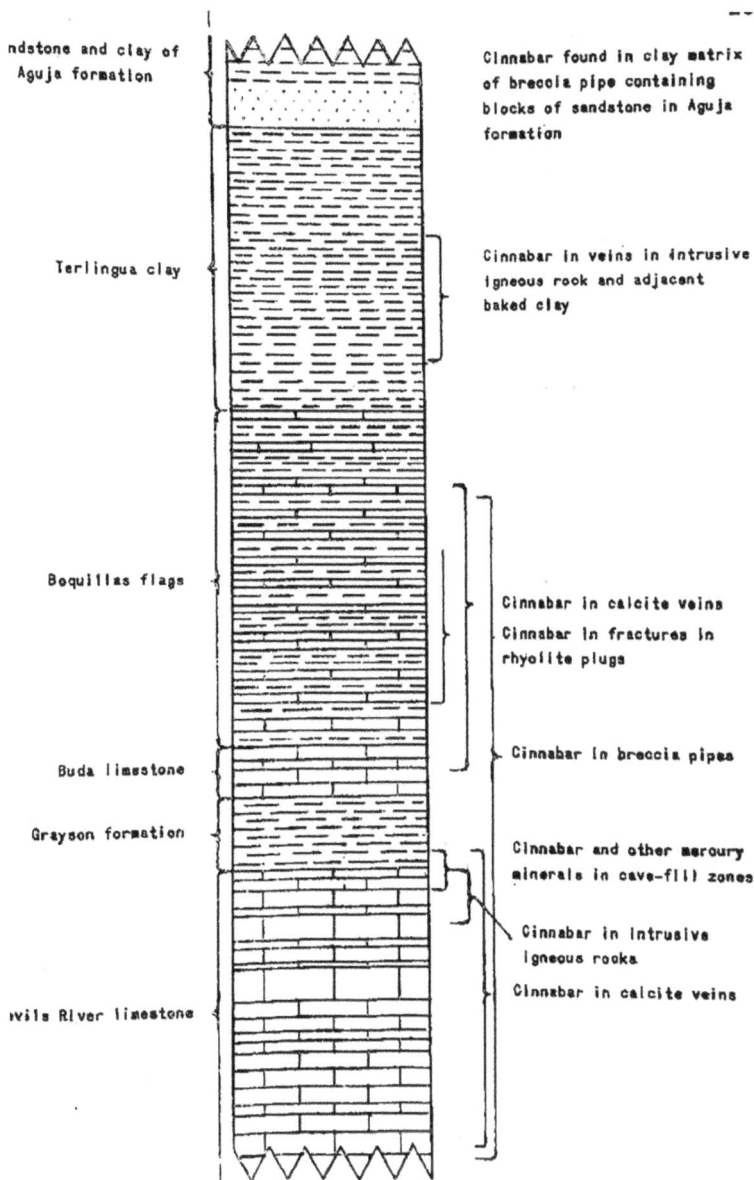

ndstone and clay of
Aguja formation

Terlingua clay

Boquillas flags

Buda limestone

Grayson formation

wils River limestone

Cinnabar found in clay matrix
of breccia pipe containing
blocks of sandstone in Aguja
formation

Cinnabar in veins in intrusive
igneous rock and adjacent
baked clay

Cinnabar in calcite veins
Cinnabar in fractures in
rhyolite plugs

Cinnabar in breccia pipes

Cinnabar and other mercury
minerals in cave-fill zones

Cinnabar in intrusive
igneous rocks
Cinnabar in calcite veins

PLATE 1

STRATIGRAPHIC SECTION SHOWING VERTICAL RANGE OF
QUICKSILVER DEPOSITS OF THE SEVERAL TYPES
THAT OCCUR IN THE TERLINGUA DISTRICT

33

CHAPTER II

SOCIO-ECONOMIC FACTORS IN MINING

Mother Nature did not have man in mind when she deposited the mercury ore in the cocky Terlingua lands. This made some of the problems of mining peculiar to the area; transportation, housing, water, fuel, labor and general mining functions were especially difficult.

For many years the only method of transportation to this isolated area was the burros and mule and the related vehicles. In the summer the rides were long, hot and tiresome. The trip from Marfa via Shafter to Terlingua would take the better of two days with a change of teams about halfway.[1] The coach or hack carried a variety of cargo. The Anglo housewives, especially, anticipated the arrival of the stage or wagon from town. Generally, a portion of the return load would be fresh fruits, vegetables and meats as well as mail from the far stretches of the nation.[2]

---

[1] C. A. Hawley, *Life Along the Border* (Spokane; Shaw and Border Company, 1955), pp. 18-22.

[2] Mr. and Mrs. Harry Fovargue, interview, April 26, 1959.

Teams of mules instead of horses were used on the hacks and wagons because the mules were able to withstand the heat, hard pulls, rough rocky trails, and constant use better than the horses.[3] The regular freight wagons were very sturdily made as their loads of bottled mercury were very heavy. A Studebaker wagon, the choice of the freighters in the mining area, would haul three or four tons of mercury at a load.[4] The freight rates were one half cent a pound, regardless of the cargo.[5] A round trip from the mines to Marfa and back would usually take about two weeks.

The mail hack made a break at the Pearl Jackson ranch. One rider would come from Marfa to this ranch, and exchange loads with the rider from Terlingua, and then each would return to his respective starting place.[6]

The route from Marfa to the mines was nearly as lonesome as a person could stand. There were no stopping places for refreshments; therefore, when the wagons left Terlingua going in with a load, the driver had to carry his provisions and stop wherever night caught him. The regular passenger hack would stop at a house about half way

---

[3] Hawley, op.cit., pp. 18-21

[4] Ibid., p. 27. A flask of mercury gross weight is one hundred pounds, seventy-six pounds of mercury and twenty-four pounds for the flask. The flasks were made of wrought iron with a threaded type metal plug that was securely tightened before shipment was made. A flask would hold about two quarts of liquid mercury. Ibid., p. 119.

[5] Fovargue, interview, April 26, 1959.

[6] Macario Hinojos, interview, March 7, 1959. Hinojos drove the mail hack for a period of two or three years.

between the two places.[7] As late as 1933, the loneliness of the area was described as

. . . in the Big Bend, the beginning of the Rockies, where are seen mountain peaks rising 9,000 feet, pure cold springs one degree above freezing and hot springs, abrupt canyons 2,000 feet deep, where a cattleman can call his ranch as empire, where one can motor 50 to 100 miles and the only person passed is a lonesome ranger, a miner, or a United States immigration officer.[8]

One of the necessities of the mines was dynamite. On one trip down to the mine a large shipment of this explosive was the cargo. The freighters did not mind hauling the explosive because it was easily handled, not too highly explosive. The caps were highly explosive and therefore the caps and dynamite were not hauled together. A woman alighted from a stage one trip to learn that five thousand caps had been under her seat on the trip. When she learned of the incident a few days later, she nearly fainted.[9]

As stopping places were few, canvas bags of water were carried for drinking purposes. The water in the bags was pleasantly cool and wet.[10] Small waterholes were occasionally found, but usually the water was not clean for drinking purposes.

---

[7] Hawley, *op.cit.*, p. 19.

[8] F. A. Hueber, "Quicksilver Industry of Big Bend Revives with New Demand for Mercury," *San Antonio Express*, 1933.

[9] *Ibid.*, pp. 19-30

[10] *Ibid.*, p. 21

As the outside world became more mobilized, the transportation at Terlingua gradually began to change. The route was then changed from Marfa to Terlingua to Alpine to Terlingua. The road had several steep pulls, and during rainy weather the truck would bog down and be on the road several days. Some of the first mechanized transportation was the three trucks purchased by the Chisos Mining Company about February 1, 1915. Felix McGaughy, at present (1960) County Judge of Brewster County, drove one of the trucks to Terlingua.[11] In late 1916 or early 1917, Jim Wade of Alpine began to haul the freight with a five-ton truck going from Alpine to the mining camp. He was the first one to do commercial hauling from the mines to town and back.[12] The wagons and teams were used partially for freighting up to 1930.[13]

The remoteness of Terlingua was also evident in the crude homes in which the Mexican laborers lived. The temporary dwellings were not any sort of protection from the elements, especially the sun of the summer. The native plants were used as pales or shades thatched together for a sun or wind break.[14] Remarks in conversation intimated

---

[11] *Alpine Avalanche*, February 4, 1915.

[12] Fovargue, interview, April 26, 1959.

[13] Stuart McGregor, "Mines in Rugged Big Bend Country Keep Texas on Map as Producer of Quicksilver," *Dallas News*, November 10, 1930. J. C. Walker, husband of the writer, has childhood recollections of seeing the freight wagons loaded and on their way to the mines. He lived several blocks due north of a railroad crossing on the Terlingua Road. Mr. Walker was born in 1923.

[14] Benjamin F. Hill, *The Terlingua Quicksilver Deposit, Brewster County, Texas* (Bulletin of University of Texas, No. 15, The University of Texas Mineral Survey Bulletin NO. 4, October, 1902), p. 11.

that heavy corrugated pasteboard, small pieces of sheet iron, odd pieces of lumber or similar materials were used. The ruins of some adobe and rock huts are standing at the present time to substantiate the facts of a one-time small industrial mining camp. Some of the Mexicans made small dugouts in the mountainside for their abodes. These homes, if some of the hovels could be called homes, were in direct contrast to the elaborately furnished home of Howard E. Perry.[15] The Anglo[16] people were provided respectable homes; a hotel of a fashion was also maintained for visitors to mines or area, while the Mexican had to provide his own home.

The dwellings of the Mexicans did not have stoves, beds, couch, chairs and tables, or any of the other household needs one would more or less expect to find in a house. All the cooking was done out of doors over an open fire that served as the stove. The rest of the furnishings were improvised in the same manner.[17]

Necessity can provide nearly anything one wants – if one wants it badly enough. One Christmas season the Anglo women wanted to decorate for a party. They did not have any way of securing the materials to the wall. Finally, someone hit upon the idea of using chewed gum.

---

[15] Perry was the owner of the Chisos Mining Company and will be discussed later in this study.

[16] The term "Anglo" is used in this study with reference to people other than those of Latin extraction.

[17] Hawley, *op. cit.*, p. 54.

One entire day was spent to chew the gum before the festive touch could be made.[18]

A very real need that necessity could not provide was water. The Anglos were allowed all the water they needed, while the Mexicans were closely rationed.[19] Periodically samples of the water in the earthen reservoir, which had been filled following the summer rains, were sent to El Paso for analysis. The result was usually "chemically pure" although the body of a donkey was found on one occasion when the water level was very low. Industrial needs of water were nil as the "dry" process of vapor reduction was used. The major portion of the water was hauled by mule teams and a tank wagon from Cigar Springs and emptied into a large galvanized steel tank near the store for the purpose of dispensing. The Mexican people had punch cards that had to be presented when they went after their water.[20]

The shortage of water affected the inhabitants only, while the fuel shortage was a real consideration of the mining staff as well as the people. The retorts and Scott furnaces were fueled by wood. [21] Mexicans were under

---

[18] Mrs. Harry Fovargue, April 26, 1959.

[19] Each adult was allowed one bucket of water a day and each child one-half bucket. The water was alkali, and it took about three weeks to get a person's system adjusted to the water. Hinojos, interview, March 7, 1959. During extremely hot dry periods of the summer, the Mexicans were allowed only two buckets of water a day. Hawley, *op. cit.*, p. 57.

[20] *Ibid.*, pp. 57-59; Hill, op. cit., p. 10.

[21] Ibid., p. 11.

contract to deliver cord wood to the mine at set rates. As the scarcity of the wood grew, so did the value increase.[22] In 1902, wood sold for $4.50 to $6.00 a cord,[23] and in 1927, a cord brought $10.00.[24]

The other production expenses had not advanced in relation to the higher prices of mercury. Common laborers were drawing the same pay in 1927 that they were in 1902, $1.00 to $1.50 per day.[25] In 1902, a flask of mercury sold for $43.20[26] and in 1927 the quotation in June was $118.92 per flask.[27] To take a day off at one's leisure was not recommended, and actually the miner could not afford to be docked that day's pay.

Out of the meager wages, a day's pay each month was withheld for medical and bare medicinal charges. The Anglo laborer paid a dollar a month for his medical services.[28] Of course, the mine foreman, all Anglos, were paid

---

[22] J. W. Furness, "Mercury," *Mineral Resources of the United States – 1927*, I, Metals, Frank J. Katz, editor, United States Department of Commerce (Washington; United States Government Printing Office, 1930), p. 68.

[23] Hill, *op. cit.*, p. 51.

[24] Furness, *op.cit.*, p. 69.

[25] Hill, op. cit., p. 60; Furness, *op.cit.*, p. 69.

[26] R. L. Ransome, "Quicksilver," *Mineral Resources of the United States – 1917*, Part I, Metals, H. D. McCaskey, editor (Washington: United States Government Printing Office, 1921), p. 421.

[27] Furness, *op. cit.*, p. 55

[28] *Ibid.*, pp. 26, 135-37.

much better salaries, and their medical fee was nominal to say the least.

The work week was seven days. Every two weeks the shift change allowed the day shift two weeks night work, and the night shift two weeks day work. When this shift was made, all the workers were off Saturday night. The Mexican people love music, singing, and dancing. There were enough good musicians to have an enjoyable evening of dancing for all ages and classes.[29]

A day's shift in the mine was ten hours long. For many years all the underground work was done by manual force of slinging the pick and hammer, and using the shovel. The shafts were so terribly hot during the summer that the miners worked naked to the waist to try to be more comfortable.[30] As mechanization progressed, the Terlingua mines gradually took on the new mode of productions. The Mexicans were not mechanically minded or inclined and therefore the pressure for improvement was not as great as it might have been.[31]

The slow development carried over into all aspects of the locale. The development was minor for the number of years the mines were in operation. The mining companies did have phone service when the telephone lines were up.[32] Perry was a shrewd businessman and dealt

---

[29] Hawley, *op.cit.*, p. 37.

[30] *Ibid.*, p. 109.

[31] *Ibid.*, p. 136.

[32] Hinojos, interview, March 7, 1959.

accordingly with his associates. He allowed only two business holidays a year, Christmas and Cinco de Mayo.[33] For the main part, lawlessness was not widespread.

In the Terlingua area each mining camp had several hundred inhabitants who made up a small village. The only name applies to these settlements for mail purposes was *Terlingua*.[34]

---

[33] Hawley, *op.cit.*, p. 32. *Cinco de Mayo* – May 5 – a Mexican national holiday commemorating the defeat of French troops at Puebla in 1862. Joseph H. L. Schlarman, *Mexico, A Land of Volcanoes* (Milwaukee: The Bruce Publishing Company, 1950), p. 315.

[34] Hawley, *op. cit.*, pp. 22, 101.

CHAPTER III

DEVELOPMENT OF THE MINES

A very small area of the vast big bend region of Texas was endowed with the much sought-after cinnabar ore. The surveyors and local mining men made many estimates of twenty-four to six hundred square miles as to the size of the productive field.[1] The district, as estimated by Sellards and Baker, is all inclusive

---

[1] Some of the varying estimates are as follows: "The quicksilver is distributed over a fairly wide area, extending from Mariscal Mountain on the east to the Lajitas Mesa on the west and from beyond the Mexican boundary on the south to the Christmas Mountains on the north. The known east and west extension of the district is about 30 miles, and the greatest north and south extensions is at least 20 miles." E. H. Sellards and C. L. Baker, *The Geology of Texas*, II, Bulletin 3401, *Structural and Economic Geology*, (January 1, 1934), 532. "The zone of mineralization is 15 miles long from east to west, by 2 miles wide." Morris P. Kirk, "The Terlingua Quicksilver District," Mining Magazine, May, 1905, p. 441. The mineral area is 8 miles east to west and 3 miles north to south. "Brewster County Minerals," *Alpine Avalanche*, March 14, 1902, citing United States Geological Survey, Bulletin 2, University of Texas Geological Survey. The Terlingua quicksilver district covered an area 5 miles wide, north and south, and 16 miles in length, east to west. John F. Lonsdale and Ross A. Maxwell, "Big Bend Has Produced Bulk of Quicksilver," *San Angelo Standard-Times*, October 9, 1955, p. 24. A rectangular strip 15 miles long and 4 miles wide has mercury bearing ore. Benjamin F. Hill, *The Terlingua Quicksilver Deposits, Brewster County, Texas*, Bulletin, University of Texas, No. 15, University of Texas Mineral Survey, Bulletin NO. 4 (October, 1902), p. 9. The following map shows Block G-12 and G-4, the principal ore areas, on page 35. Clyde P. Ross, "Preliminary Report on the Terlingua Quicksilver District Brewster County, Texas," Sellards and Baker, *op. cit.*, p. 560.

of the area where the cinnabar has been found, not just where the closely related Terlingua mines are located. This ore is found about 104°W longitude and 29°30'N latitude.[2]

As the news reached the outside world from the lonely, barren stretches of the new found mercury ore, people began to go into the region and stake claims as close as possible to one another.[3] The first known publication concerning the quicksilver deposit was in the *Bullion*, Los Angeles, California, August, 1894; and about the same time the *Manufacturer's Record*, Baltimore, carried an article. W.P. Blake's article in *Transactions of the American Institute of Mining*, 1895, was the first scientific report about the newly discovered mercury ore.[4] George W. Wanless, Jimenez, Mexico, and agent for Rio Grande Smelting Works, and Charles Allen, Socorro, New Mexico, made an exploration of the lower Brewster County area and found the mercury ore. Wanless had heard of some Mexicans finding rich ore in the aforementioned area. Blake was informed of the discovery by James P. Chase, Socorro, New Mexico, and the two men visited the region together in August, 1894.[5]

---

[2] William P. Blake, "Cinnabar in Texas," *Transactions of the American Institute of Mining Engineers*, XV (Florida Meeting, March, 1895), 69.

[3] William Battle Phillips, "The Quicksilver Deposits of Brewster County, Texas," Economic Geology, I, 1905, 155. A standard claim is 1500 feet = 500 varas long; 600 feet = 216 varas wide. One mine claim is 20.66 acres. Trans-Pecos Abstract Office, Alpine, Texas.

[4] Phillips, "The Quicksilver Deposits . . .," p. 155.

[5] Blake, *loc. cit.*

Several individuals staked claims, set up retorts of furnaces and "closed" before operation actually had time to get started. Much of the work was started before a thorough knowledge of the amount of ore could be ascertained. The first form of refining by retort was considered a "poor man's" method of mining.[6] The first ore refined was taken from the surface of the land, hand sorted and processed in retorts.

Numerous claims were staked on which the mines were established.[7] Some of the companies that were established by 1905 were Chisos Mining Company, Marfa and Mariposa Mining Company, Terlingua Mining Company, Colquitt-Tigner Mining Company, Texas Almaden Mining Company, Big Bend Cinnabar Mining Company, and the Excelsior Mining Company. All of these mines were short-term producers as compared with the Chisos Mining Company.[8]

Early mine laws of Texas were a hindrance to out-of-state mining interests.[9] The Colquitt-Tigner Mines were

---

[6] "Terlingua," *Alpine Avalanche*, October 24, 1902.

[7] List of mine claims, Appendix III.

[8] John T. Lonsdale, director, *Texas Mineral Resources*, Bureau of Economic Geology, The University of Texas Publication, No. 4301, January 1, 1943 (Austin; University of Texas, 1946), p. 374.

[9] "Terlingua Mining Law," *Alpine Avalanche*, June 6, 1902. This article is a discussion of mining laws of Texas as reported by the Commissioner of General Land Office. "Our Mining Laws," *Alpine Avalanche*, November 23, 1900; *The Mining Laws of Texas and Tables of Magnetic Declination*, The University of Texas Mineral Survey, Bulletin 6, July, 1903, Bulletin of the University of Texas, No. 21.

involved in a Supreme Court decision that finally gave the mining interest the signal to progress with its development.[10]

The Marfa and Mariposa was the first mining company to be organized and actually start production.[11] Montroyd Sharpe had a ten-ton furnace installed in 1900 by Robert Scott[12] and a staff of mechanics.[13] In 1901, Norman, Sharpe, and Golby were owners of the Marfa and Mariposa. By the end of 1902 this mine alone had produced 9001flasks of mercury with a gross sale of $371,788 being reported for 1901 and 1902. At the reported period, the New York quotation of mercury per flask was $48.00. The claims that were worked by the Marfa and Mariposa were the ones taken up originally by Chase and Allen.[14]

The crude, simplified processes used in the mining operations gave the earliest claimants good returns on their investments. The deepest shaft was two hundred feet; but most of the ore was removed from "shallow workings, trenches, pits, etc., by windlasses, horsewhim, or gasoline hoist."[15]

---

[10] "Miners Get Mineral Lands," *Alpine Avalanche*, May 16, 1902.

[11] Morris P. Kirk, "The Terlingua Quicksilver District," *Mining Magazine*, May 1905, pp. 441-442.

[12] Robert Scott was the designer of a furnace for process mercury-bearing ore. The furnace bore his name and was used over a period of several years in the Terlingua district.

[13] Letter to *Alpine Avalanche*, June 18, 1900.

[14] Joseph Struthers, "Quicksilver," *Mineral Resources of the United States*, 1902, David T. Day, editor (Washington: Government Printing Office, 1904), p. 255.

15 Phillips, "The Quicksilver Deposit . . .," p. 160.

The fuel for the furnaces was wood selling between $5.50 and $7.00 a cord. One and a half to two cords of wood a day would serve to roast twenty to thirty tons of ore a day. A small coal deposit had been located a few miles from the mines, and some was being used at $6.00 to $6.50 a ton.[16] The Mexican helper was paid $1.00 to $1.50 a day, depending on his previous mining experience. The Mexican worked very well with an Anglo to supervise the work and prod the laborers just a little.[17]

All the supplies were wagon freighted to the mines. Marfa and Mariposa received 2,300 empty flasks in 1900 and were waiting for the completion of the furnace to resume operation.[18] The latter part of 1900, $30,000 worth of mercury was shipped to New York. During the same period, six wagon loads of machinery arrived at the Marfa and Mariposa.[19] Rich ore had been found, furnaces were being built, and possibilities of a glowing future beckoned the mine enthusiast and adventurer. Prospecting was going on everywhere in the area.[20] Scott, who had come to Terlingua to install a ten-ton Scott furnace for Marfa and Mariposa, said the quicksilver belt was the greatest on earth if indications developed as hoped for.[21] This mine produced well during its period and had a very high grade

---

[16] *Ibid.*

[17] Hill, *op. cit.*, pp. 50-51.

[18] *Alpine Avalanche*, August 3, 1900.

[19] *Ibid.*, December 14, 1900.

[20] *Ibid.*, December 28, 1900.

[21] *Ibid.*, June 8, 1900, citing *Marfa New Era*.

of ore.[22] The greatest amount produced in one month by 1910 was 432 flasks.[23] An estimate of the total production through 1909 was 40,000 flasks, each weighing seventy-five pounds, at an approximate value of $1,800,000.[24] Considering labor, fuel, and the grade ore processed, the mercury was netting between $15.00 and $17.00 per flask.[25]

This fabulous industry with all its possibilities of being able to pay good living wages worked its men for near starvation wages. The Mexican mine laborers did all the digging, ten hours a day, seven days a week, for $1.00 to $1.50 a day, depending on their ability and knowledge of mining. A few very skilled miners received $2.00 a day.[26] Terlingua had a population of 1,000 to 1,500 Mexicans and very few Anglos.[27] C. A. Hawley, one of the Anglo employees at the mine, was bookkeeper at the Marfa and Mariposa Mining Company. After talking with Clifford Dennis, mine superintendent, Hawley was assigned odd jobs from storekeeper to justice of peace of the area. His

---

[22] William B. Phillips, "Condition of the Quicksilver Industry in Texas," *Engineering and Mining Journal*, LXXXVIII (November 20, 1909), 1022.

[23] Robert Scott, "Modern Quicksilver Reduction," *Mining and Scientific Press*, C (January 22, 1910), 164.

[24] Phillips, "Condition of Quicksilver . . .," p. 1022.

[25] *Ibid.*, p. 1024.

[26] C. A. Hawley, *Life Along the Border* (Spokane: Shaw and Border Company, 1955), p. 25.

[27] *Ibid.* Terlingua was first located at the Marfa and Mariposa Mining Company and later moving to the Chisos Mining Company where it is today. *Ibid.*, pp. 22, 101.

salary was $100 a month plus free house rent and store goods at cost. Hawley began his work, including duties of justice of peace, without a gun in 1906 and remained in the area until 1913.[28]

The mineral deposits had been nearly exhausted; California Hill had yielded her rich ores for several years; and they, too, were gone. A great deal of native mercury had been found; but the miners felt it was time to stop working a vein when they found native mercury in quantities. When native mercury is found, it is usually at the bottom of a deposit.[29]

The working crew had been reduced to an exploration crew and the mine's activities began to wane. All the old workings were re-examined in the hope that new deposits could be found. The tunnels were lighted only by candles as the mine did not have electricity. This method of exploration was dangerous as the candles gave very little light. While on one of these exploratory ventures, Dennis fell in an air circulation shaft and was slightly injured. Little activity was carries on after that. For a period of some twelve to fourteen years this mining company had been a strong producer.[30] By 1910, the only mine operating in the Terlingua district was the Chisos Mine.[31]

---

[28] *Ibid.*, pp. 9-23.

[29] *Ibid.*, p. 96.

[30] *Ibid.*, pp. 96-100.

[31] H. D. McCaskey, "Quicksilver," *Mineral Resources of the United States*, 1914, Part I, Metals, George O. Smith, editor (Washington: Government Printing Office, 1916), p. 329. The operation of the Chisos Mine is discussed elsewhere in this study.

While the Marfa and Mariposa was still at its peak, James Norman took and exhibition of stones the size of small watermelons, to the St. Louis World Fair in 1904. These stones could be broken open, having a hollow center, with a beautiful crystalized lining, containing some native quicksilver.[32] The mining fever spread rapidly over the rough terrain of the lower portion of Brewster County. Mining companies were organized, furnaces or retorts installed, but little or no production was made. A situation involving the mercury-bearing area was the dispute that grew out of a county survey made by Joe Moss in 1900 concerning surveys "40, 41, and 59 in Block G12."[33] The dispute was settled in 1902 accepting the "Joe Moss" line as the correct survey.[34] Lindheim and Dewees were organizing a mining company, and soon found themselves involved in the "Joe Moss" line dispute. W. W. Turney had filed charges against Lindheim and Dewees for recovery in payment for some ore removed from the boundary of the adjoining properties. The court proceedings took two or three years to settle. Before the final judgement was rendered, Lindheim and Dewees sold their holdings to Marfa and Mariposa Mining Company; Turney's group sold to the Terlingua Mining Company. The parties buying the disputed lands accepted the survey and agreed to honor the line [35].

---

[32] *Ibid.*, p. 96.

[33] "The Disputed Line," *Alpine Avalanche*, July 27, 1900.

[34] "'Joe Moss Line' Established," *Alpine Avalanche*, July 18, 1902.

[35] *Ibid.*

Mine holdings of John Goughran and Charley Hess established about 1898 were sold to Pelican State Oil and Mineral Company of Shreveport I August, 1901.[36] The property of Hines and Lafarelle on Section 34, Excelsior Mine, had also been sold to the Pelican Oil Company.[37] The company then became known as Colquitt-Tigner.[38] In 1902, Colquitt installed a ten-ton Scott furnace and made plans to treat the ore on the Excelsior Mine claims. Only surface operations were being worked in veins eight inches to three feet wide and some pockets of ore deposits.[39] A report of 1905 states that the furnace had been idle for the past two years; however, a rich ore supply had been discovered and production started again.[40] Various parcels of land were added from time to time although the production was seemingly negligible. The Colquitt-Tigner was not mentioned as a producing mine for several years.[41]

Another of the early organized mines was the Terlingua Mining Company. The company installed a forty-five-ton

---

[36] "History of Quicksilver Mining in Big Bend Area Shows Many Years' Activity," *Alpine Avalanche*, Sixtieth Anniversary Edition, September 14, 1951, p. 44.

[37] "Mining Deal," *Alpine Avalanche*, August 23, 1901.

[38] "History of Quicksilver . . . ," *Alpine Avalanche*, September 14, 1951.

[39] Struthers, *op.cit.*, p. 255.

[40] F. W. Horton, "Quicksilver," *Mineral Resources of the United States*, 1905, David T. Day, editor (Washington: Government Printing Office, 1906), p. 396.

[41] Phillips, "Condition of Quicksilver . . .," p. 1022; Charles A. Dinsmore, "Quicksilver Deposits of Brewster County, Texas," *The Mining World*, XXXI (October 30, 1909), 877.

Scott furnace and was ready to start production. The organization had a capital of $500,000 and staked several claims in various sections. Most of the ore was near the surface, and at that time the deepest shaft was seventy feet and very few shafts had been dug.[42] In 1903, the evaluation of the Terlingua Mining Company was $35,025.[43] In a very short time the mining company ceased operations.[44]

Premature investments in the Terlingua district were evident by the number of mining companies that files claims, set up companies, put up capital but did not go any further. Some of the claims were abandoned, and others were absorbed by the large production companies. Red Cloud Mining Company,[45] Western Cinnabar Mining Company,[46] Excelsior Mining Company,[47] United States Quicksilver Mining Company,[48] Brewster County Mining

---

[42] *Alpine Avalanche*, October 24, 1902.

[43] *Ibid.*, July 31, 1903.

[44] Dinsmore, *op. cit.*, p. 877; Phillips, "Condition of Quicksilver . . , p. 1022.

[45] H.W. McGuirk to Alexander Boynton, Trustee for Red Cloud Mining Company, Deed Records, Brewster County Clerk's Office, Alpine, Texas, VIII, 425-427.

[46] F. E. Gillett to Western Cinnabar Mining Company, Quit Claim Deed, *ibid.*, X, 310.

[47] *Alpine Avalanche*, August 23, 1901.

[48] John Gaughran to United States Quicksilver Mining Company, Bond and Lease, Deed Records, Office of the County Clerk, Brewster County, Alpine, Texas, VIII, 47.

Company,[49] Texas Cinnabar Mining Company,[50] and Lone Star Mining Company[51] were some of the very minor or non-producing companies. The McKinney Brothers and J. M. Parker staked claims just south of the Chisos Mines but disposed of the property in a short time.[52]

Two mines close to the Terlingua district, but in a district of their own geologically, are those of the Big Bend Cinnabar Mining Company and Texas Almaden Mining Company located on the north and south sides of Study Butte.[53] Claims originally filed on by W. L. Study, Survey 216, Block G-4, are quicksilver claims – La Fortuna, El Porvenir, El Pueblo, Tyrone, Juaniata, and El Ojo – a one-third of two-thirds interest conveyed by Federico Billalba and Teburcio de la Rosa to C. Alarcon and by D. Alarcon to Santos Terrazas, transferred to Serapio Sanches, Ojinaga, Chihuahua, Mexico, as quit claim deeds filed May 1, 1903. An option and contract by W. L. Study, Federico Billalba, Tiburcio de la Rosa and D. Alarcon with M. P. Kirk, E. G. Gleim and H. S. Gleim was made June 16, 1902.[54] Sanchez sold 588 shares each to Fred Smith and Herbert

---

[49] Brewster County Mining Company to Charles G. Jester, Trustee, Note Holder Brewster County Mining Company, *ibid.*, XXXI, 420.

[50] H. W. McGuirk to Terlingua Cinnabar Mining Company, *ibid.*, VIII, 330.

[51] Charles Scheiner, *et al.* to Lone Star Mining Company, *ibid.*, X, 50; Phillips, "Conditions of the Quick-silver . . ," p. 1023.

[52] McKinney Brothers, *et al.*, Terlingua Mining Company, Deed Records, Office of the County Clerk, Brewster County, Alpine, Texas, V, 268.

[53] Phillips, "Quicksilver Deposits . . .," p. 157.

[54] Santos Terrazas, Quit Claim Deed to Serapio Sanchez, Deed Records, Office of the County Clerk, Brewster County, Alpine, Texas, VIII, 638.

Bryant, in the Big Bend Cinnabar Mining Company, in November, 1903.[55] In May, 1905, Federico Billalba bought five thousand shares at one dollar each from the Big Bend Cinnabar Mining Company. E. G. Gleim was president and C. Spencer Gregg was secretary of the company.[56] The company called a director's meeting in late 1927 to consider leasing, with option to buy, to John H. Leavell and W. D. Burcham. The mine had not operated since 1920 because it was unprofitable, shafts were flooded with water, and the company was in debt with no funds to pay off. The board decided to call a shareholders' meeting and further discuss the lease proposition and elect new company officers.[57] The Big Bend Cinnabar Mining Company did their ore extraction in underground shafts. In 1918, the shafts were reported to be about 1300 feet deep.[58] The Big Bend Cinnabar Mining Company had a fifty-ton Scott furnace to use for its reduction of ores.[59] This mine and its sister mine, Texas Almaden Mine, were worked rather spasmodically because of water seepage.[60] For a period of thirty-eight months including the year 1915, the Big Bend

[55] Serrapio [sic] Sanchez to Fred Smith, ibid., IX, 114; Serapio Sanchez to Herbert Bryant, ibid., IX, 115.

[56] Big Bend Cinnabar Mining Company to Federico Billalba, ibid., XXIII, 195.

[57] Big Bend Cinnabar Mining Company to Leavell et al., Corporate Minutes, ibid., LIX, 589.

[58] W. D. Burcham, "Quicksilver in Texas in 1918" Engineering and Mining Journal, CVII (January 18, 1919), 145.

[59] Phillips, "Condition of the Quicksilver . . . ," p. 1022.

[60] Big Bend Cinnabar Mining Company to and with Leavell et al., Corporate Minutes: Directors Meeting, Marfa, Texas, Deed Records, Office of the County Clerk, Brewster County, Alpine, Texas, LIX, 589.

Cinnabar Mining Company produced 5,550 flasks of mercury.[61] Big Bend Cinnabar Mining Company had a strong water seepage at a level of one hundred fifty feet between 1918 and 1920.[62]

The Texas Almaden Mining Company was chartered in November, 1905, with a capital stock of $200,000.[63] This mine was owned principally by Sanger Brothers, Dallas, and some of their associates.[64] The activities of this mine were small and in 1927, Burcham and Leavell leased it along with the Big Bend Cinnabar Mining Company and referred to the properties as Brewster Quicksilver Consolidated.[65]

The small mines opened and closed at various times, hoping that a new vein of rich ore could be found. The "get-rich-quick" dream died shortly for most mine enthusiasts. "Mines are made; -- not discovered."[66]

---

[61] W. D. Burcham, interview, Alpine, Texas, March 1, 1959.

[62] Burcham, interview, March 1, 1959; Big Bend Cinnabar Mining Company to and with Leavell *et al.*, Corporate Minutes; Directors Meeting, Marfa, Texas, Deed Records, Office of the County Clerk, Brewster County, Alpine, Texas, LIX, 589.

[63] Charter Texas Almaden Mining Company, *ibid.*, IX, 626. Spain has a mine named Almaden, that has reportedly produced for twenty-eight centuries. Almaden means "the great mine" which the one in Spain certainly is. The Texas namesake did not live up to its name. F. A. Hueber, "Quicksilver Industry of Big Bend Revives with New Demand for Mercury," *San Antonio Express*, 1933; Sellards and Baker, *op.cit*, pp. 516-518.

[64] Charter Texas Almaden Mining Company, *ibid.*, IX, 626.

[65] Paul M. Tyler, "Quicksilver," Mineral Resources, 1930, Part I, Metals, O. E. Kiessling, editor (Washington: Government Printing Office, 1933), p. 48.

[66] E. A. Waldron, "Quicksilver and Mineral Development," *Alpine Avalanche*, January 1, 1920.

CHAPTER IV

CHISOS MINES

The fact that mines are made, not found, is evidenced by the development of the claims at the Chisos location. The buried treasure in the awesome surrounding made Perry a very wealthy man.[1] This seeming wasteland near the Rio Grande did not encourage men to live or to work there. Robert T. Hill's description of the land in 1900 still holds in 1960. Speaking of the plants, he said, "They [the plains] are covered by spiteful, repulsive vegetation, the chief feature of which is the ocotillo, a plant with small green leaves on long and slender stalks that reach above a substructure of lechugilla, cactus, sotol and other thorny plants, like serpents rising from a Hindu juggler's carpet."[2] The Terlingua quicksilver district is in this type of locale. The Chisos Mine was located near the center of the mining activities.

---

[1] Paul M. Tyler, "Quicksilver," *Mineral Resources of the United States, 1928,* Part I, Metals, Frank J. Katz, editor (Washington: Government Printing Office, 1931), 278.

[2] Robert T. Hill, "Running the Canons of the Rio Grande," *The Century Magazine,* LXI (January, 1901), 373. Hill was on a United States Geological Survey, and his descriptions of the region from Presidio, Texas, down the Rio Grande through Santa Helena Canyon are very interesting accounts.

During the productive days the sole owner and operator of the Chisos Mine was Howard E. Perry, a northern businessman. The rugged, arid parcel of land that was first secured by Perry was purchased in 1887 for $5,760 and consisted of four and one-half sections of Blocks G-12 and G-4.[3]

Several additional sections were purchased from time to time.[4] In the original survey of Block G-4, sections 295 and 296 were put in the wrong places, and later, the section numbers and location were corrected. In this correction it was Perry's very good fortune that the most productive part of the Chisos Mine lay there. Otherwise, Perry might never have been in the mining industry.[5]

The Chisos Mine continued its operation longer than any other mine in the district. By 1917, only one mine in the United States, the New Idria in California, surpassed

---

[3] Wilbur A. Reeves to Howard E. Perry, Warranty Deed, June 10, 1887, Deeds Records, Office of the County Clerk, Brewster County, Alpine, Texas, II, 585. Several persons have related an incident of Perry's first land holding as being acquired as partial payment of a debt to Henderson Shoe Company, St. Louis, Missouri. Henderson was Mrs. Perry's father. Mr. Perry was in the lumber business in Cleveland, Ohio. This land has not, however, been recorded in the deed records at Alpine, Texas. One writer states, "Ed Nevill, local ranchman, insisted that he saw Mr. Perry buy two sections of land at an auction in front of the courthouse in Alpine and that he paid $150 cash for them." Virginia Madison, *The Big Bend Country* (Albuquerque: University of New Mexico Press, 1955), p. 183. This sale has not been recorded in the courthouse.

[4] Appendix No. 2.

[5] Deed Records, Office of the County Clerk, Brewster County, Alpine, Texas, II, 585-88; III, pp. 62, 67-70.

its production. The shafts at the Chisos were also the deepest in the given area.[6]

The ore of the Chisos Mines occurred "along a strong curved fault zone with a generally nearly east to west trend."[7] The fault curves to the north; the downward slope, generally speaking, is nearly vertical; the principal ore-bearing part of the area has a "downthrow" that does not exceed sixty feet, notwithstanding the possibility of a like horizontal element in the shifting of the faulted area. The three shafts that were operating and the portion being worked had an opening 2,500 feet long along the fault line and some 750 feet deep, with the shafts about 50 feet apart vertically. The rocks uncovered from the surface and going down into the ground have thin-layered covering of yellowish shale with the related Austin Chalk, most likely not exceeding a depth of fifty feet; five hundred feet thickness of Eagle Ford formation, one hundred feet of Buda limestone, between seventy-five and one hundred feet of Del Rio clay, and a thickness of Edwards limestone. The Upper Cretaceous formations are the Austin chalk and the Eagle Ford, and the Lower Cretaceous formations are Buda limestone, Del Rio clay, and Edwards limestone. The chief source of the ore at the Chisos Mine was in the Buda limestone; however, large amounts were removed above the 550-foot level of the overlying Eagle Ford; and the first part of 1917 some ore was found at a depth of 700 feet in

---

[6] F. L. Ransome, "Quicksilver," *Mineral Resources, 1917* – Part I, H. D. McCaskey, editor (Washington: Government Printing Office, 1921), p. 422.

[7] *Ibid.*

the Edwards limestone. Little exploration had been done at this level because it was below water level.

The mineral-bearing ore from the surface to a depth of six hundred feet occurred at intervals along the explored area of the faulted line, particularly where there are several cleavages or a function of the cross cleavages. However, the richest ore of the mine was below the six-hundred-foot level in a chimney extending downward to the seven-hundred-fifty-foot level with a diameter ranging from seventy-five to one hundred feet. The chimney extends into the Del Rio clays with the shattered Buda limestone due likely to a depressed faulting in the clay shale, probably resulting from the collapsing of a "solution chamber" in the weighty Edwards limestone. This dropped ore mass of chaotic breccia is principally Buda limestone, with cinnabar, native quicksilver, free sulphur, and less prominent minerals. The major production of the Chisos Mine for a period of three or four years was nearly in its entirety from this rich ore formation.[8]

The mining activities of the Chisos Mines evolved principally around one person, Howard E. Perry. The physical, material, and cultural development relied entirely on the whims of this man. He was sole investor, along with his creditors, in the Chisos Mining enterprise.

Perry controlled the town of Terlingua, which settlement consisted of a movie house, drug store and soda fountain, church, commissary, bank of a fashion, a school, homes of varying degrees, and the necessary garages and

_____

[8] *Ibid.*

repair shops for machinery and other needs. The now vacant, rundown school building is still referred to as "Perry's School." On a small hill overlooking all the townsite is a large two-story house that Perry and his wife used when they were in Terlingua. The old house gives a slight impression of an old castle perched high above other buildings and surroundings to give the "master" a better view of his vast domain. Perry could have been referred to as the king of Terlingua." He was definitely ruler over the Anglos and Mexicans who worked for him. One of his contemporaries wrote,

> Mr. Howard E. Perry, was a successful Chicago business man and possessed very little of the milk of human kindness. He never learned a word of Spanish, and never manifested any interest in the Mexican people as fellow human creatures.[9]

> The association Perry had with the Mexicans was purely business. Perry was very secretive about all his business, private and commercial.[10]

Perry was not in Terlingua very much of the time. Daily mining reports were sent to him in Chicago or wherever he was.[11] Mr. Perry did not want anyone knowing about his

---

[9] Hawley, *op.cit.*, p. 101.

[10] Madison, *loc. cit.*

[11] Hawley, *op.ci.*, p. 102

mining activities as to the income, amount of ore mined, flasks of mercury bottled and sold, or the selling price.[12]

Perry had only "Mr. Perry's" interest in mind when dealing with his Mexican laborers. The men's wages were $1.00 to $1.50 a day for a ten-hour work day.[13] Perry owned all the businesses in Terlingua and consequently nearly all of the money paid out in wages returned to him. Part of the laborers were paid in American money, and some in Mexican money; a few were paid in both. The exchange value varied from time to time, but at this time it was two pesos for one dollar.[14]

The education of the children in the area was scattered and scant. The village of Terlingua did not provide educational facilities for its children. In February, 1929, County Judge C. D. Wood corresponded with R. L. Cartledge in regard to a school.

While it is yet a little early, I wish you would take up with Mr. Perry the construction of a good school building

---

[12] Madison, *op. cit.*, pp. 183, 191-193; Harry Fovargue, Alpine, Texas, April 26, 1959. Mr. Fovargue stated that Mr. Perry did not want anyone to know anything about his business. That was strictly for him to know. Therefore, very little actual information in regard to production, sales, and financial reports is available.

[13] Hawley, *op.cit.*, p. 25.

[14] Ibid., p. 131. Virginia Madison relates that regardless of the value of the dollar and peso exchange, Perry would use the rate of two to one. At times the exchange would be five or six to one. The people were forbidden to trade elsewhere, and to encourage their trading in Terlingua the store items were supposed to be in relation to the money exchange " . . . but the plan was spoiled by neighbors who purchased pesos at the market price and spent them in his store, thus flooding the market." *Op. cit.*, p. 189.

at Terlingua this summer. Something with three or more rooms having in view a centralized school for all the surrounding country. This is my idea and I believe it would work out – by paying someone to bring in scholars every school day from different directions.[15]

Later a school was built. Perry provided for that, too, with certain strings attached.

The main obstacle to overcome down there in school matters is that Mr. Perry admittedly, frankly and certainly controls the School Trustees and two are on his payroll – and the mine runs the school to suit Mr. Perry. One result of building the school house at Terlingua, it was hoped, would be that there would be only one school in that area, and therefore the Mexicans would move to Chisos and work in the mine, as labor continuously became more scarce and less tractable. The other idea, and which was perhaps more important, was that a better school would keep in the camp the children of the white employees, so the wives would not move to town and disrupt the family life of the American workmen. Another factor affecting the trustees is that Mr. Perry charged such a rental for the Terlingua School House they are afraid there is not enough money to have other schools more than four and a half months a year, or to pay adequate rentals for other school buildings. Mr. Perry demanded a thousand dollars a year but takes or accepts about $800 a year rental, or

---

[15] Letter, C. D. Wood to R. L. Cartledge, Alpine, Texas, February 26, 1929.

did the last I heard. Although he threatened each year to raise it.[16]

The homes furnished the laborers, Anglo and Mexican, were of many kinds. The "king" himself had his "castle" on the hill. This home, though seldom lived in, was lavishly furnished. Carpeted floors, beautiful ornate light fixtures of all sorts, very fine furniture and all of the accompanying features filled the Perry home. The Anglo families had small but serviceable homes, while the Mexicans lived in dwellings from dirt floored adobe huts to caves in the hills.[17] The hardy, tolerant, easy going Mexican people slaved to make Perry an extremely wealthy man.

In the earliest days of mining, the ore was taken from the top of surface of the land. The cinnabar ore was light to dark red in color. Had the surface rock at Terlingua been of the "impervious Terlingua Marl instead of relatively brittle, much jointed and hence relatively pervious Eagle Ford flags" the cinnabar ore most likely would not have yet been found. Dr. J. A. Udden encouraged the miners to go much deeper to get to the best ore after some mining

---

[16] Madison, *op. cit.*, pp. 191-192, citing a letter from Tom Skaggs to Mr. Townsend, April 11, 1936. Mr. Skaggs was not employed by Mr. Perry; therefore, he felt at liberty to criticize the manner in which the Terlingua school situation was handled. The school house and church were built in the early 1910's. Hinojos, interview. Hawley speaks of the lack of a school or church. He left Terlingua in 1913. Soon thereafter the buildings were constructed. Hawley, op. cit., pp. 108-110.

[17] Fovargue, interview and Hinojos, interview.

had already been done. The faulting of the area happened after the ore was deposited, making the ore harder to find.[18]

A portion of the ore found in the early years of mining on the property of the Chisos Mining Company was assayed to be 25 per cent metallic quicksilver. This ore lay in veins of shale. At that time, 1905, very little work as to underground mining had been done.[19] However, by 1917, Shaft Number 9 was 750 feet deep, and water in small amounts had been encountered. Shaft Number 8 was 800 feet deep, but no water was in evidence. The water of Shaft 9 was pumped out into storage tanks, and water was no longer rationed.[20]

As previously mentioned, the principal ores of the Chisos mining are cinnabar, three peculiar oxychlorides – terlinguaite, eglestonite, montroydite, -- found only in the Terlingua mining district, and native or "free" mercury.[21] At one location in the Terlingua area are some caverns where very rich grade ore has been found in piles of the floor of the caverns. Some of the pieces of ore weighed up to fifty pounds and assayed at 30 per cent mercury. One of the caverns located was forty-five feet deep from the

---

[18] Ross, op.cit., p. 552.

[19] Morris P. Kirk, "The Terlingua Quicksilver District," Mining Magazine, May, 1905, p. 442.

[20] Fovargue, Interview.

[21] Ransome, op. cit., p. 402; William B. Phillips, "The Quicksilver Deposits of Brewster County, Texas," Economic Geology, I, 1905, p. 155.

surface, twenty-five feet high, sixty feet long and forty feet wide.[22]

In the earliest days of operation in the Terlingua district, the crudest and simplest form of mining and recovering of the mercury was applied. Large dumps or piles of ores were collected before actual work was begun. Vertical retorts were used at first in the removal of the mercury from the ores.[23]

A Scott furnace was soon introduced in the area. The furnaces were spoken of as being "10 ton," "20 ton," and "50 ton." The numeral designated was the number of tons the furnace could process in a twenty-four-hour period.[24] The furnace was economical for processing the lower grades of ores. The rich ores with which the Chisos Mining Company was endowed were successfully and profitably processed in the retort. The initial installation cost of the furnace was a great disadvantage to their early users in setting up new operation activities. Other adverse qualities of the furnace were the amount of soot in the condenser saturated in mercury, health danger, poor condensing procedure, especially on low grade ore, brick absorption of mercury vapors by condensers. Periodically the brick condensers had to be dismantled, crushed, and fired in the furnace to retrieve as much of the mercury as possible. On a per ton basis of a furnace, each ton capacity would cost

---

[22] Phillips, "The Quicksilver Deposits . . .," p. 158.

[23] *Alpine Avalanche*, December 24, 1902.

[24] Fovargue, interview, Alpine, Texas, April 26, 1959.

about $1000.00.[25] The smallest tonnage furnace was the ten ton. By 1909, a possible daily reduction capacity of 170 tons was in evidence in the Terlingua district. However, because of overestimation of the amount of ore present, only two furnaces with a daily capacity of thirty tons were in operation. The idle furnaces were due partly to the immature knowledge of the field, lack of constant labor and shortage of firewood.[26]The Scott furnace used wood as its fuel. During a twenty-four-hour period, a twenty-ton furnace would use about one and one-half to two cords of wood at $10.00 a cord. Much later some of the furnaces were changed and diesel engines were installed. The fuel was hauled to the mine from Marathon.[27]

Huttner and Scott had built a furnace to reduce cinnabar ores on a twenty-four- hour working basis. The furnace, referred to as monitor or ironclad, had a hexagon shaped base with three fire-places around it to burn wood. Above the fire-places the shaft was round with a vapor chamber at the top, which carried the fumes to the condenser through iron pipes. The furnace stood eleven feet high with a width of six feet. The untreated ore entered the furnace at the top by a cup and cone arrangement. As a new load was put into the top, a hermetical seal was used to hold the temperature of the ore below that was

---

[25] E. Bryant Thornhill, "Wet Method of Mercury Extraction," Mining and Scientific Press, CX (June 5, 1915), p. 873.

[26] Phillips, "Condition of Quicksilver . . . ," p. 1022.

[27] J. W. Furness, "Mercury," Mineral Resources of the United States – 1927, I, Metals, Frank J. Katz, editor, U. S. Department of Commerce (Washington: United States Government Printing Office, 1930), p. 68.

still being processed. As untreated ore entered the top, exhausted or processed ore was removed from the bottom of the furnace. Air heat bypassing through clay pipes in the condenser served for combustion and also helped keep the amount of soot to a minimum. The ore chamber was 27 feet high and 25 inches wide and 11½ feet long. The inside of the furnace resembles two sets of stairways standing vertical to each other with the angles fitting in between one another. There are two sets of the vertical stairways in the furnace, and the ore tumbled from step to step, back and forth, on its way through the furnace. The heat and hot air going through the furnace carried the vapor from the furnace to the condenser. As the vapor reached the condenser, it was near the boiling point.[28]

The boiling point of mercury is 357°C or 675°F.[29] Mercury begins to give off a vapor at 40°C.[30] The vapor is a definite health hazard as it salivated the glands of the mouth resulting in the loss of teeth.[31] Mercury vapors are sweet smelling and the very earliest reducing was done out of doors in order to protect the health of the men.[32]

---

[28] Arthur H. Hiorns, *Principles of Metallurgy* (London: McMillan and Co., Ltd., 1941), pp. 362-63.

[29] Clifford A. Hampel, *Rare Metals Handbook* (New York: Reinhold Publishing Corporation, 1954), p. 629.

[30] Hiorns, *op. cit.*, p. 362.

[31] Georgius Agricola, *De Re Metallica*, trans. Herbert Clark Hoover and Lou Henry Hoover (New York: Dover Publications, inc., 1950), p. 426.

[32] Agricola, op. cit., p. 427. Agricola has described the processing of cinnabar ore in the sixteenth century in this fashion. Vinegar and salt were used to clean the "free" mercury that was found. The mercury was poured into a leather or

In condensing the mercury vapor to a liquid form, iron condensers, brick chambers or thin glass may be used in the process. The brick chambers are not satisfactory for long periods of use as they gradually get hot and cool very slowly. Water has been used in some mines in the condensation process, although it was never used in the Terlingua area. In the monitor furnace, the vapors leave the "take off" at 350°-400°C and upon entering the first condenser the temperature is 200°C, and by the time the vapor reaches the brick condenser the temperature is 40°C, passing through the glass condenser at 14°C before passing into the chimney. A mercury-saturated soot is removed from the condenser and roasted with fresh ore. Wood ashes were used to brighten some of the mercury that was in the sooty substance.[33]

After this silver liquid metal had been bottled into flasks, it was ready for numerous commercial uses. The use most commonly thought of is the thermometer. Mercury freezes at -38.87°C to -40°C,[34] and as stated before

---

canvas bag, then squeezed or forced through the bag and caught in pans or pots. The ore was reduced by using two pots; one resembling a sauce pan, the other similar to a jug. The jugs were filled with crushed ore, sealed with moss, inverted into the sauce pan and joined together with lute. Seven-hundred of these pots were prepared at one time for processing. After the pots were sealed together, they were placed in sand, earth or ashes to the top of the small pan. Then the fire was built and reduction proceeded. The jugs and pans were made of a very good potters clay to prevent the escaping of mercury vapors. This was the most common way of reducing ore; however, four other methods were used and discussed in the book.

[33] Hiorns, *op cit.*, p. 363.

[34] Hampel, *op. cit.*, p. 629; Sellards and Baker, *op. cit.*, p. 511. De Mille's *Strategic Minerals*, page 314, gives the melting point, which is nearly

boils at 357°C. One of the uses that has determined to a great extent the rise and decline in price per flask is fulminate, used to detonate high explosives. During war time the price immediately goes up. The following statement and table will demonstrate the price scale as well as five the number of flasks produced per year and the value of mercury for the period of time including the years 1899 to 1917.[35] The Chisos Mine did not begin marketing mercury until 1903.[36]

There were 4 producing mines in 1917 all in Brewster County. The ore treated amounted to 28,242 short tons and yielded 10,791 flasks of quicksilver, valued at $1,136,508.00. In 1916 25,945 tons yielded 6,306 flasks, valued at $793,862.00. The production of Texas from the beginning of operations in the Terlingua district to date is given in Table III. [37]

Other uses would number about three thousand.[38]

---

synonymous with freezing as "38.8°C" omitting the minus which was either an oversight of the author or a typographical error.

[35] Ransome, *op. cit.*, p. 421.

[36] Stuart McGregor, "Mines in the Rugged Big Bend Country Keep Texas on Map as Producer of Quicksilver," *Dallas News*, November 10, 1930.

[37] Ransome, *op. cit.*, p. 421.

[38] De Mille, *op. cit.*, p. 314. Some of the applications besides thermometers and munitions are drugs, chemicals, corrosive sublimate, vermillion, mercuric mixtures (solutions), acetic acid, phthalic acid, phthalic anhydride, into which mercury does not enter itself, rabbit fur treatment to be used in making men's felt hats, amalgamation of gold and silver ores, thermostats, antifouling marine paint for ship bottoms, solution to prevent boiler scale, cosmetics, dental amalgam, and aid in floating lights in houses to give a suggestive list

The Chisos Mine, during its life span, produced its fair share of mercury from a monetary standpoint as well as the quantity. The following table gives a comparative analysis of some California mines and three Texas mines. The Texas mines most used in the comparison were the Chisos, Mariposa, and Study Butte. The table is principally to show a comparison of labor production.[39] As to the exact production of the Chisos during its forty years of yielding, the true amount will never be known. Perry felt that was his business and not for public knowledge.[40]

---

of varied uses of mercury. Ransome, op. cit., pp. 383-85. The Basket -Makers of the area used the cinnabar ore in relation to their dead. "Some evidence has been found in caves, of the bones of the deceased having been divested of flesh and painted with a red substance, the basis of which was cinnabar." H. Conger Jones, "Mercury Mines, Treasure House of Texas' Big Bend, Long Neglected," *San Antonio Express*, March 18, 1934.

[39] Ransome, op. cit., p. 378. In comparison, the amount of ore treated by the Texas mines was small; however, the number of flasks of mercury was comparatively greater. The ore of the Texas mines was a richer ore than that of the California mines.

[40] Furness, op. cit., p. 68.

TABLE III

QUICKSILVER PRODUCED
IN TEXAS, 1899-1917[41]

Year	Quantity (flasks)	Price per flask	Value
1899	1,000	$ 47.70	$ 47,700
1900	1,800	44.94	80,892
1901	2,932	48.46	142,085
1902	5,319	43.20	229,781
1903	5,029	45.29	227,763
1904	5,336	43.50	232,116
1905	4,723	36.22	171,067
1906	4,761	39.50	188,060
1907	3,686	39.60	145,966

[41] Ransome, *op. cit.*, p. 421

1908	2,382	44.17	105,213
1909	4,188	45.45	190,345
1910	3,320	46.51	154,413
1911	2,326	46.01	107,019
1912	1,990	42.05	83,680
1913	2,750	40.23	110,633
1914	3,144	49.05	154,213
1915	4,417	85.80	378,979
1916	6,306	126.89	793,862
1917	10,791	105.32	1,136,508

TABLE IV

RELATIONS OF QUICKSILVER PRODUCTION TO LABOR IN CALIFORNIA AND TEXAS[42]

STATE	Ore treated (short tons)	Quicksilver produced flasks	Tenor won	Average Number of men employed	Man days of work	Quantity of ore treated per man (short ton)	Quantity of ore treated per man days (short ton)	Quicksilver obtained per man annually (flasks)	Quicksilver obtained per man day (flasks)
California									
All mines	235,786	23,938	0.38	950	310,500	248.1	0.75	25.1	0.077
8 leading mines	203,449	18,581	0.34	715	233,689	322.8	0.99	26.1	0.079
Texas									
3 mines	26,802	10,530	1.47	253	84,795	105.9	0.32	41.6	0.124

[42] Ransome, *op.cit.*, p. 378

CHAPTER V

PASSING OF THE MINING PERIOD

When a person drives into the present Terlingua, a paved highway leads nearly to the old haunts of quicksilver mining days. As the mercury ore became scarce, mining companies diminished by consolidation or liquidation. One unusual transaction started in 1924 with North Texas Company selling its property to M. B. Whitlock *et al.*[1] The group that bought the property organized the Rainbow Mining Company.[2] The mine was located adjoining the Chisos Mining Company on the east. A story is told of the Rainbow group's desire to sell out to the Chisos for something near $9,000.00. Perry of the Chisos Mine, did not take up the offer. Instead, he extended his tunnels adjoining the Rainbow property. Evidently this report was true. In 1930, M. B. Whitlock *et al.* filed suit against the Chisos Mining Company for trespassing. The suit netted the Rainbow concern $75,000.[3]

---

[1] North Texas Company to M. B. Whitlock *et al.*, Warranty Deed, Deed Records, Office of the County Clerk, Brewster County, Alpine, Texas, LI, 52.

[2] *Ibid.*

[3] M. B. Whitlock *et al.*, vs Chisos Mining Company, District Court Minutes, Office of the County Clerk, Brewster County, Alpine, Texas, IV, 309-311.

During the 1930's, the Chisos Mining Company bought out the Rainbow Mining Company.

In the early 1920's E. A. Waldron began to buy up mining properties. One of these properties purchases was Colquitt-Tigner.[4] Another former producer, Marfa and Mariposa, was in the process of being sold by various persons. W. Van Sickle sold Sections 41 and 59 to Texas-Mariposa Mining Company in 1928.[5] Four years later, 1932, the Texas-Mariposa Mining Company was sold to the Chisos Mining Company.[6] The mines of Texas-Mariposa were worked again under the new owner, the Chisos Company. A payroll record for one week for the Mariposa mine showed the wages ranging from $1.40 a day to $2.00 a day. A total of eighteen men were working 58 hours cumulatively for a total sum of $1383.[7] The production of the mines began to diminish greatly in the late 1930's. A statement of the Chisos Mining Company for June, 1939, shows a total expense of $6,017.10 in contrast to a low production of thirteen flasks of quicksilver at $40 each.[8]

Although the mines were beginning to be non-productive, Perry still tried to control the area. In July, 1938, while

---

[4] Colquitt-Tigner to E. A. Waldron, Deed Records, Office of the County Clerk, Brewster County, Alpine, Texas, XLIX, 581; LI, 220.

[5] Van Sickle to Texas-Mariposa Mining Company, *ibid.*, LXXIV, 533.

[6] Texas-Mariposa Mining Company to The Chisos Mining Company, *ibid.*, LXXXIII, 567-68.

[7] Memo sheet, Mariposa Payroll, 4/4/36, possession of the writer.

[8] Memo sheet, statement of The Chisos Mining Company, June, 1939, possession of the writer.

Perry was in Chicago, he contacted Cartledge, his Terlingua business manager, to be sure to have Bill Burcham return to Terlingua and get the vote of the area. Burcham, a mining man, was Perry's choice. Perry said, "We must win --."[9]

For forty years the Chisos Mines were stripped of their mercury-bearing ores for Perry. The total amount of flasks and monetary income is not exactly known. The production from all the mines over the years is reported to be 150,000 flasks. The Chisos Mines are estimated to have produced about two thirds of the total amount.[10] The income made a fortune for Perry; but he took bankruptcy in 1943 to sever himself from the mines.[11] Perry, while traveling by train to Florida, on December 6, 1944, passed away.[12] Brown and Roat of Houston at the present own the mines. A small amount of work was done by the new company; but the shafts were soon lost to underground waters.[13]

On going into the area, the traveler will see abandoned adobe, rock, or frame buildings still standing, minus roofs, windows, doors, or floors. The old "castle" on the hill is as stripped as any building around it. Waste dumps form

---

[9] Letter from Howard E. Perry to Robert Cartledge, Chicago, July 12, 1938.

[10] Robert G. Yates and George A. Thompson, *Geology and Quicksilver Deposit of the Terlingua District Texas*, Geological Survey Professional Paper 312 (Washington: United States Government Printing Office, 1959), p. 100.

[11] Chisos Mining Company, Bankruptcy Proceedings, Deed Records, Office of the County Clerk, Brewster County, Alpine, Texas, CIII, 193.

[12] Barry Scobee, "Terlingua – Town That Mining Ghosted," *San Angelo Standard-Times*, June 8, 1958.

[13] *Ibid.*

small mounds in the nearby valley. A few Mexicans still live in Terlingua, hoping that somehow the mines will open again.[14]

To the east, at the Study Butte area, the Big Bend Cinnabar Mining Company sold to Texas Mercury Company in 1940, for a reported $10,000.[15]

Compared to the mines of Spain, Terlingua was a very short-lived mining area.[16] Many people have tried in the past few years to revive the mines. The ore is economically depleted. The mines passed as the frontier of the area passed. The lonesome ruggedness is still evident, but for the present its colorful, prosperous days have also passed.

The quicksilver mines might be likened to a century plant. The treasure or bloom was there for many years; but once it put forth its blossom, it died quickly.

---

[14] Macario, Hinojos, interview, Terlingua, Texas, March 7, 1959.

[15] Big Bend Cinnabar Mining Company to Texas Mercury Company, Deed Records, Office of the County Clerk, Brewster County, Alpine, Texas, CV, 214.

[16] F. L. Ransome, "Quicksilver," Mineral Resources of the United States, 1917, Part I, Metals, H. D. McCaskey, editor (Washington: Government Printing Office, 1921), p. 394.

ESSAY ON SOURCE MATERIAL

Gathering the material for this study has been a great challenge. It has been similar to a treasure hunt to find needed information, with one clue leading to another. The initial research was done in the *Alpine Avalanche* dating from January 1, 1900. Various articles quoted other references; in turn those articles were obtained if possible. *The Terlingua Quicksilver Deposits, Brewster County, Texas*, Texas University Mineral Survey Bulletin 4, 1902, was borrowed from the University of Texas. This report also made reference to other sources. Because much of the information was published around 1900, the volumes could not be sent on loan. Thermo-Fax or Copease reproductions of the materials were obtained from several sources throughout the country. *Mineral Resources of the United States*, 1917, has a very extensive bibliography on quicksilver, several items of which articles refer to the Terlingua Mines.

If one knows nothing of the ruggedness of the lower part of Brewster County, one should read Robert T. Hill's account in *Century Magazine*, January, 1901, entitled "Running the Cañons of the Rio Grande." The main difference between now and then is the paved highway from Alpine and Marathon to the area. The Mexicans still live their simple life with few modern conveniences. Interviewing a few people that lived and worked in the area during its peak helped a great deal. The writer visited the area to view the present condition and observe the remains of the physical facilities.

BIBLIOGRAPHY

A. PRIMARY SOURCES

1. Books

Hawley, C. A. *Life Along the Border*. Spokane: Shaw and Border Company, 1955.

> Personal experiences of Hawley while employed at the mines in the Terlingua area. For the most part very reliable.

2. Publications of Government Agencies and Scientific Societies

Furness, J. W. "Quicksilver," *Mineral Resources of the United States, 1926*, Part I, Metals, Frank J. Katz, editor. Washington: United States Government Printing Office, 1929.

> Report of 1926 including Waldron's work.

Furness, J. W. "Quicksilver," *Mineral Resources of the United States, 1927*, Part I, Metals. United States Department of Commerce. Washington: United States Government Printing Office, 1930.

Annual report on the minerals of the United States. Reliable statistics pertaining to all phases of minerals, mining, processing, and sales.

Hill, Benjamin F. *The Terlinqua Quicksilver Deposit, Brewster County, Texas.* Bulletin 15.

University of Texas Mineral Survey Bulletin 4, October, 1902.

Very thorough geological report with the addition of some useful general material.

Hillebrand, W. F., and W. F. Schaller. *The Mercury Minerals from Terlingua*, Texas. United

States Geological Survey, Bulletin 405, 1909. Thorough description of the various mercury-bearing ores of the Terlingua area.

Horton, F. W. "Quicksilver," *Mineral Resources of the United States*, 1905, David T. Day, editor. Washington: Government Printing Office, 1906.

Discussion of the development of the Terlingua mining area by 1905 with a very concise account of the organized mines.

McCaskey, H. D. "Quicksilver," *Mineral Resources of the United States, 1914*, Part I, Metals,

George Otis Smith, editor. Washington: Government Printing Office, 1916.

Brief production report to 1910 and activity from 1910 to 1914.

The Mining Laws of Texas and Tables of Magnetic Declination. The University of Texas

Mineral Survey, Bulletin 6, July, 1903, Bulletin of the University of Texas, No. 21.

Perry, C. C., and Arthur Schott. *United States and Mexican Boundary Survey.* Geological Reports. Washington: Cornelius Wendell, 1857.

Survey report of the border area between the United States and Mexico, including some information on plants, animals, and native population.

Ransome, F. L. "Quicksilver," *Mineral Resources of the United States, 1917,* Part I, Metals,

H. D. McCaskey, editor. Washington: Government Printing Office, 1921.

Annual report giving a great deal of pertinent material, such as production charts, sales, labor yield, and general conditions of the Texas mines up to 1917.

Ross, Clyde P. "Preliminary Report on the Terlingua Quicksilver District, Brewster County, Texas," *The Geology of Texas,* II, *Structural and Economic Geology,* E. H. Sellards and C. L. Baker, editors. Bulletin 3401, January, 1934.

Concise geological and mining report which is a good bibliographical aid.

Struthers, Joseph. "Quicksilver," *Mineral Resources of the United States, 1902*, David T. Day, editor. Washington: Government Printing Office, 1904.

Brief report of early operation and production of mines established by 1902.

Tyler, Paul M. "Quicksilver," *Mineral Resources of the United States, 1928*, Part I, Metals,

Frank J. Katz, editor. Washington: Government Printing Office, 1931.

Brief report of activities at Terlingua in 1928.

Tyler, Paul M. "Quicksilver," *Mineral Resources of the United States, 1930*, Part I, Metals, O.

E. Kiessling, editor. Washington: Government Printing Office, 1933.

Consolidation of Study Butte mining interest.

Udden, J. A. *The Anticlinical Theory as Applied to Some Quicksilver Deposits.* University of

Texas Bulletin 1822, 1918.

Report on anticlinical theory of quicksilver deposits, with brief reference to the Terlingua area.

Udden, J. A. *A Sketch of the Geology of the Chisos Country, Brewster County, Texas*. Bulletin

93, Science Series No. II. Austin: University of Texas, April 15, 1907.

Technical account of geology of the Chisos area.

3. Periodicals

Blake, William P. "Cinnabar in Texas," *Transactions of the American Institute of Mining*

Engineers, XV (1896), 68-76.

Early report given at a Florida mining meeting as to the initial disclosures and verification of the presence of mercury ores in the Terlingua area.

Burcham, W. D. "Quicksilver in Texas in 1918," *Engineering and Mining Journal*, CVII

(January 18, 1919), 145-146.

Brief report on the various mines and sales relating to the quicksilver industry at Terlingua in 1918.

Dinsmore, Charles A. "Quicksilver Deposits of Brewster County, Texas," *Mining World*, XXI

(October 30, 1909), 877-878.

General discussion of the mining industry in Terlingua at that time.

Hill, Robert T. "Running the Cañons of the Rio Grande," *Century Magazine*, LXI

(January, 1901), 371-387.

Vivid description of the area in 1900 that is still good today.

Kirk, Morris P. "The Terlingua Quicksilver District," *Mining Magazine*, May, 1905, pp. 440-443.

Early account of the development of the mines, showing a few pictures of the industry in 1905 during its infant years.

Moses, A. J. "Eglestonite, Terlinguaite, Montroydite, New Mercury Minerals from Terlingua, Texas," American Journal of Science, XVI (September, 1903), 253-263.

Very technical report on mineral ores peculiar to the Terlingua quicksilver mining area.

Phillips, William Battle. "Condition of the Quicksilver Industry in Texas," *The Engineering and Mining Journal*, LXXXVIII (November 20, 1909), 1020-1024.

Discussion of the immaturity and lack of knowledge in the development of the quicksilver industry of Terlingua. A detailed report for the period, 1909.

Scott, Robert. "Modern Quicksilver Reduction," *Mining and Scientific Press*, C (January 22, 1910), 164.

Defensive rebuttal about the Scott furnace and its satisfactory operation in various mining areas in the United States.

Strauss, Lester W. "Modern Quicksilver Reduction," *Mining and Scientific Press*, March 19, 1910, p. 431.

Brief discussion of types of reduction used in quicksilver mining.

Thornhill, E. Bryant. "Wet Method of Mercury Extraction," *Mining and Scientific Press*, CX

(June 5, 1915), 873-874.

Explanation of the types of extraction of mercury.

Turner, H. W. "The Quicksilver Deposits," *Economic Geology*, I (1905), 265-281.

Geological report of area as the area was developed by 1905.

Turner, H. W. "The Terlingua Quicksilver Mining District, Brewster County, Texas," *Mining and Scientific Press*, June 21, 1900, p. 64.

4. Documents

Office of the County clerk, Brewster County, Alpine, Texas. Commissioners Court Minutes, I.

Office of the County Clerk, Brewster County, Alpine, Texas, Deed Records, Vols. II, III, IV, V, VII, VIII, IX,

X, XX, XXIII, XXXI, XXIV, XL, XLII, XLIII, XLIX, LI, LVIII, LIX, LX, LXIII, LXXIV, LXXXIII, XCII, CIII, CV, CVII.

Office of the County Clerk, Brewster County, Alpine, Texas. District Court Minutes, IV.

Office of the County Clerk, Brewster County, Alpine, Texas. Patent Records, Vol. II.

Miscellaneous memoranda relating to operation of the Chisos Mines in Possession of the writer.

Trans-Pecos Abstract Office, Alpine, Texas. List of mining claims and their location.

United States Court. Western Division of Texas. Decree, May 13, 1901. Office of the County Clerk, Brewster County, Alpine, Texas, Deed Records, VIII, 164-165.

5. *Newspapers*

Alpine Avalanche, January 1, 1900 – January 1, 1941.

Numerous articles in this publication gave pertinent information as well as references to other sources. A great deal of material from the issues of the first two or three years was used. This publication is not indexed, and the work is very tedious. Issues from a few years before this time would have been of great help in leads to the earliest interest in the area.

6. Interviews and Letters

Burcham, W. D., and Mrs. W. D. Burcham. Alpine, Texas, March 1, 1959. Interview.

> Helpful explanation of mining processes and condition of the mines by a mining engineer.

Fovargue, Harry, and Mrs. Harry Fovargue. Alpine, Texas, April 26, 1959. Interview.

> Very informative, as Mr. Fovargue served as a bookkeeper for the Chisos Mining Company for several years.

Hinojos, Macario. Terlingua, Texas, March 7, 1959. Interview.

> Accurate old gentleman still living in Terlingua who speaks and understands English well and remembers many facts about the quicksilver days of Terlingua.

Williams, O. W. Letter to his children, Alpine, Texas, March 18, 1902.

> Manuscript in the possession of Mrs. F. G. Walker, Alpine, Texas.

Wood, C. D. to Robert Cartledge, Alpine, Texas, February 26, 1928.

> Letter in possession of writer.

B. SECONDARY MATERIALS

---

1. Books

Agricola, Georgius. De Re Metallica. Translated by Herbert Clark

> Hoover and Lou Henry Hoover. New York: Dover Publications, Inc., 1950.
>
> Very early (1550) book on metallurgy. In the footnotes supplied by Mr. Hoover, a brief history of mercury is written to show how long its services have been used by man. Agricola gave some pertinent information in his book that is valid today. His discussion on mercury is interesting and thorough.

Cooke, Philip St. George, William Henry Chase Whiting, and Francois Xavier Aubry. Exploring Southwestern Trails, 1846-1854, Ralph P. Bieber, editor. Glendale: The Arthur H. Clark Company, 1938.

> Map, p. 387, showing trails explored for possible trade between San Antonio and El Paso.

DeMille, John B. *Strategic Minerals*. New York: McGraw Hill Company, Inc., 1947.

> Brief but informative discussion on properties and uses of mercury.

Hampel, Clifford A. Rare Metals Handbook. New York: Reinhold Publishing Corp., 1954.

Discussion of rare metals, of which quicksilver is considered to be one.

Hiorns, Arthur H. Principles of Metallurgy. London: Macmillan and Co., Ltd., 1914.

Thorough understandable discussion on metallurgy contemporary with the period of operation of the Terlingua mines.

Jackson, W. Turrentine. *Wagon Roads West*. Berkeley: University of California Press, 1952.

Study made for possible roads, including one between El Paso and San Antonio.

Map, p. 38, shows trails used by various groups between 1846 and 1869.

Madison, Virginia. *The Big Bend Country*. Albuquerque: University of New Mexico Press, 1955.

Nonfiction book of the Big Bend Country based on historical facts but lacking accuracy in some details.

Richardson, Rupert Norval, and Carl Coke Rister. *The Greater Southwest*. Glendale: The Arthur H. Clark Company, 1934.

Good general history of the Southwest with some mention of closely related area happenings.

Rister, Carl Coke. The Southwestern Frontier, 1865-1881. Cleveland: The Arthur H. Clark Company, 1928.

Background information on the early development of the Southwestern frontier.

Schlarman, Joseph H. L. Mexico – A Land of Volcanoes. Milwaukee: The Bruce Publishing Company, 1950.

Narrative, colorful, history of Mexico.

Schulz, Ellen. *Texas Wild Flowers*. Chicago: Laidlaw Brothers, 1928.

Study of cacti giving common and scientific names, description and some interesting facts about the plant.

Thrall, H. S. *History of Texas*. St. Louis: N. D. Thompson & Co., 1879.

Interesting and useful narrative with a social slant.

2. Dictionary

Van Nostrand's Scientific Encyclopedia. Princeton: D. Van Nostrand Company, Inc., 1958.

Broad coverage of all the science given principally in dictionary form.

3. Publications of Government Agencies

Gannett, Henry. A Gazetteer of Texas. United States Geological Survey, Bulletin 224, Series F, Geography 36. Washington: Government Printing Office, 1916.

Detailed survey giving place names and corners of various land areas, contemporary with the period of production of the Terlingua mines.

Lonsdale, John. *Texas Mineral Resources*, Bureau of Economic Geology, The University of Texas Publication, NO. 4301. Austin: University of Texas, 1946.

Brief geology report with numerous references and helpful as a bibliographical aid.

Sellards, E. H., and C. L. Baker. *The Geology of Texas*, Vol. II, *Structural and Economic Geology*. The University of Texas Bulletin, No. 3401 (January 1, 1934). Austin: University of Texas.

Brief geology report with numerous references and helpful as a bibliographical aid.

"Texas Prehistory – Indians," *Texas Almanac*, 1847-1957. Dallas: Dallas Morning News, 1956-1957.

Brief summary of Indians that inhabited Texas before it was developed.

Yates, Robert G., and George A. Thompson. Geology and Quicksilver Deposits of the Terlingua District, Texas. Geological Survey, Professional Paper, 312. Washington: United States Government Printing Office, 1959.

Up-to-date report and future possibilities of the Terlingua area.

4. Periodicals

Gibson, Freda. "Local Place Names," West Texas Historical and Scientific Society, Bulletin 21, I, December 1, 1926.

> Cleverly written article about place names and the folklore behind them.

Witte, Adolph H. "Terlingua Creek," *Bulletin of the Texas Archeological and Paleontological Society*, XV (September, 1934), 108-110.

> Based on relics, and includes information on the vegetation of the area.

5. Unpublished Materials

Cepeda, Paulina. "History of Terlingua," term paper, Sul Ross State College [1956], citing unpublished manuscript of A. W. Fulcher, 1949.

> Recollections of happenings and stories told about the area. A very simple report.

6. Newspapers

The Alpine Avalanche, Sixtieth Anniversary Edition, September 14, 1951.

> Articles in this edition summarize development of quicksilver mines.

Hueber, F. A. "Quicksilver Industry of Big Bend Revives with New Demand for Mercury," *San Antonio Express*, 1933.

General exposition of quicksilver and use up to
that period.

Jones, H. Conger. "Mercury Mines, Treasure House
of Texas' Big Bend, Long Neglected," *San Antonio
Express*, March 18, 1934.

Brief resume of world, as well as Texas, quicksilver,
with special emphasis on Terlingua quicksilver district.

Lonsdale, John T., and Ross Maxwell. "Big Bend Has
Produced Bulk of Quicksilver," *San Angelo Standard-
Times*, October 9, 1955.

Synopsis of the life of Terlingua in a rather pictur-
esque description.

McGregor, Stuart. "Mines in Rugged Big Bend Country
Keep Texas on Map as Producer of Quicksilver,
Dallas News, November 10, 1930.

Feature story on the mines in Terlingua as they
were in 1930, with a brief history included.

APPENDIX I

The Mexican people were poor, as well as isolated, and had to use the native plants to help supplement their way of living. The Mexicans and pioneers made use of the native plants[1] in many ways as fiber, food, medicine, building material, drink, stock food and fuel. Many of these uses still continue, especially among the poor people during droughts and financially hard times.

The *Yucca treculeana*, also known as Spanish Dagger, Don Quixote's Lance, and Pita, is one of the useful plants because of the leaf and flower. While the leaf is green, it can be held over a fire, softening the tissue making it possible to peel strands of fiber from the leaf. The strands are then tied together and serve the purpose of rope or twine. The blossom can be gathered and cooked as cabbage or be made into a tasty pickle. The needle-point of the Yucca has been used as a medicine of great importance as it can be used on animal and man after either has been bitten by a rattlesnake. The Apache used the leaf as material for weaving baskets.

---

[1] The best description and classification of these plants is Ellen D. Schulz, *Texas Wild Flowers* (Chicago: Laidlaw Brothers, 1928), 47-257.

The *Yucca tenuistyla*, known as Beargrass, Yucca or Adam's Needles, has been used for hanging meats and a sewing twine on sacks or heavy coarse materials. This fiber is stronger and finer than sisal hemp. The plant contains saponin, which serves as cleaner for wool and as a hair shampoo.

Dasylirioin texanum, commonly called Dasylirion, Saw Yucca or Sotol, is used for the leaf and the cabbage-like head. The leaf is sometimes used to thatch roofs of the Mexican and Indian dwelling, whether adobe or a lean-to-room made of tree limbs. The head has been used as a stock food after the outer part has been burned off. The heads can be baked, buried, and soured to make a strong alcoholic beverage know as *sotol* or *sotol mescal*. In early oil exploration test, the plant was used as fuel for firing boilers.

Agave americana, also called Century Plant, Agave, Maguey, American Aloe, contains a sap in the leaves that make an intoxicating drink that Mexicans have used for a long time. This is a stronger drink than whiskey. In a fermented stage, it has been given the name *pulque*, and in a distilled form is called *tequila*. Century plants have been grown commercially in Mexico. The leaf can also be heated and crushed to make a solution for cough syrups and poultices.

Agave Lechuguilla, known as Lechuguilla or Lechequilla, have fiber that may be used for brushes, matting, coarse twine, and rope; also, the trunk of the plant has been used in making soap. The Lechuguilla is also a source for making mescal, a favorite drink. The Apache Indians and Mexicans baked the head in the ground and ate it.

Berberis trifoliolata, commonly known as Agarita, Agrito, Chaparral Berry, and Wild Currant, have berries that make delicious jellies and wines.

The *Prosopis pubescens* or *Strombo-Carpa pubescens*, called Screw Bean, or Tornillo, have bean pods that are boiled to make a sort of molasses. When the bean is dry, it can be ground and used as flour. The flour can be soaked in water to make a nutritious drink. The wood of the plant is a very good fuel. The branches are strong enough to make fences or huts for the Mexicans.

Prosopis juliflora, well-known as Mesquite, Mezquite, Algarobe, was one of the mainstays of the Mexican people as food for human and stock consumption, fence posts, fuel and medicine. The early Indians, as well as the Mexicans, used the beans raw or dried. The people would chew the ripe pods in the raw form. When the beans were dried, a flour was made by grinding them on a *metate*. A mild form of beer can be produced by boiling the flour and then letting the mixture ferment. The dried leaves can be made into a tea that served as a medicine. Domestic animals, particularly horses and mules, enjoy the beans, especially in dry seasons.

Larrea mexicana, better known as Cresote Bush, Greasewood, Cabonadera, Gobernadora, Hediondilla, is used for its medicinal purposes and antiseptic effects. Twigs and leaves of the plant are steeped to make a solution especially beneficial in treating cuts, bruises, and sores whether for man or beast, and was exceptionally good to heal saddle gall.

Fouquiera spendens, also called Candlewood, Jacob's Staff, Ocotillo, Coach-whip cactus, serves as material for making a living fence and as starting wood for camp fires.

Opuntia lindheineri, commonly known as Prickly Pear or Nopal, though slightly distasteful, has been used by the poor Mexicans, who eat the fruit of the cactus raw. The families of moderate means make jellies and candies of the fruit. The young and tender leaves of the prickly pear are edible after being cooked. As the leaves or joints grow older, they can be boiled and used as a poultice. The prickly pear, like many of the other cacti, is also a source of good alcohol. During the dry years, the people with stock burn the spines or thorns, so that the stock can eat the plant.

Lophophora *williamsii*, known also as Cactus, Dry Whisky, Devil's Root, Mescal Button, Sacred Mushroom, and Peyote, was known to have a narcotic aspect by the Indians. The effects are very similar to those of opium. Some Indians and Mexicans were superstitious about the spineless plant and believed it was capable of warding off evil and bad luck.

These plants are some of the most common used by the inhabitants of the Terlingua area.

APPENDIX II

E arly reports were made to the effect that Perry obtained his initial holding through a debt to his father-in-law, Henderson Shoe Company, St. Louis, Missouri. The records of land holdings in his name while mining, do not bear this fact out. The grantee is Howard E. Perry in each instance; therefore, only the grantor, acreage, block number, and date are given.

Grantor	Acreage	Block No.	Date
Reeves, Wilbur A.	4-1/2 sections	G-12 & G-4	June 10, 1887
Sherriff Sale [1]	1 section	G-12	November 22, 1899
Holland, J. R. and wife	72 63/100 acres	G-12	May 15, 1908
Olguin, Ursulo	640 acres	G-4	March 22, 1910

---

[1] $30.75. This was the cheapest land that Perry bought in the Terlingua area.

Lafarelle, Jas.	640 acres	G-4	April 13, 1910
Cartledge, Eugene	640 acres	G-4	January 13, 1917
Hernandez, Vidal	2 sec. 406.13 acres	G-4	February 27, 1917
Cartledge, Eugene	640 acres	G-4	November 20, 1917
Garcia, Julian	3 sections	G-4	September 26, 1918
Van Sickle, W.	1 section	G-4	October 17, 1929
Cartledge, R. W.	1 section	G-4	February 12, 1930

APPENDIX III

A representative list of the names of mine claims filed for some of the mining companies of the Terlingua area are as follows:[1]

MARFA AND MARIPOSA MINING COMPANY		
Margaret C	Georgia L	Last Chance
Keystone	Jose	Rothchild
Al Reed	Waldron	McKinley
Dan White	Rosaura	Robo
I.C.L.	Diamond	Mariposa
Lindheim	Zoe	Santiago
Emma G	Raymond Zinky	Richardson
H.G.S.	Bonanza	Monte Cristo
Pilkington	Hoover	Agage
Opal	Topaz	Comanche
Apache	Gard, A. J.	Hope
Don Juan	Croesus	Linton
Canyon	Kimberly	Jimachee
Mahattic	Maggie	Minnie
D. & G.	Sample 1-8	Curtis

[1] Trans-Pecos Abstract Company, Alpine, Texas.

Columbia	Lindberg	F. W. Oakes, Jr.
Duncan 1-8	Kathleen E	Excelsior
Abra Chico	Abra Grande	Murray 1-4
Brown 1-3	Sharpe 1-2	Nelson 1-2
Juan 1-7	Hughes 1-2	Bruces
(M.P.P.) Sirdar	Little	Hulen
White Heather	Regina	Koll-I-Moor
Spirit of St. Louis	Caledona	Sentimenta
CHISOS MINING COMPANY		
Franklin	Packard	Essex
Super-Six	M.P. 1-3; 8	Reo Fifth
Flying Cloud	Wolverine	Speedwagon
Bacine	Ford	Ned 1-9
F. A. Waldron	Howard E. Perr	Terlingua 1-7
Link	64; 1-4	Pontiac
Cadillac	Buick	Oakland
G. H. Smith 1-7		

9 7 8 1 6 3 7 6 5 3 1 0 4